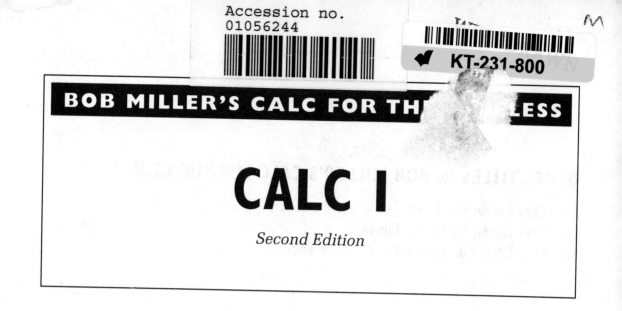

BOB MILLER'S CALC FOR THE CLUELESS

CALC I

Second Edition

1

OTHER TITLES IN BOB MILLER'S CALC FOR THE CLUELESS

Bob Miller's Calc for the Clueless: Calc II, Second Edition
Bob Miller's Calc for the Clueless: Calc III
Bob Miller's Calc for the Clueless: Precalc, Second Edition

CALC I

Second Edition

Robert Miller

Mathematics Department
City College of New York

McGraw-Hill

New York   San Francisco   Washington, D.C.   Auckland   Bogotá
Caracas   Lisbon   London   Madrid   Mexico City   Milan
Montreal   New Delhi   San Juan   Singapore
Sydney   Tokyo   Toronto

*To my wife, Marlene, I dedicate this book and anything else
I ever do. I love you. I love you! I LOVE YOU!!!!!*

Library of Congress Cataloging-in-Publication Data

Miller, Robert, date
    Calc I / Robert Miller. — 2nd ed.
        p.   cm. — (Bob Miller's calc for the clueless)
    Rev. ed. of: Bob Miller's calc I helper. ©1991.
    Includes index.
    ISBN 0-07-043408-5 (paper)
    I. Miller, Robert, date   Calc I helper.   II. Title.
    III. Series: Miller, Robert, date   Calc for the clueless.
QA303.M68512   1998
515—dc21   97-43438
                                                   CIP

BOB MILLER'S CALC FOR THE CLUELESS: CALC I

    4 5 6 7 8 9 10 11 12 13 14 15 16 17 18 19 20 DOC DOC 9 0 2 1 0 9

ISBN 0-07-043408-5

Sponsoring Editor: Barbara Gilson
Production Supervisor: Clara Stanley
Editing Supervisor: Maureen Walker
Project Supervision: North Market Street Graphics
Photo: Eric Miller

McGraw-Hill

A Division of The McGraw·Hill Companies

ACKNOWLEDGMENTS

I have many people to thank.

I would like to thank my wife Marlene, who makes life worth living.

I thank the two most wonderful children in the world, Sheryl and Eric, for being themselves.

I would like to thank my brother Jerry for all his encouragement and for arranging to have my nonprofessional editions printed.

I would like to thank Bernice Rothstein of the City College of New York and Sy Solomon at Middlesex County Community College for allowing my books to be sold in their book stores and for their kindness and encouragement.

I would like to thank Dr. Robert Urbanski, chairman of the math department at Middlesex, first for his encouragement and second for recommending my books to his students because the students found them valuable.

I thank Bill Summers of the CCNY audiovisual department for his help on this and other endeavors.

Next I would like to thank the backbones of three schools, their secretaries: Hazel Spencer of Miami of Ohio, Libby Alam and Efua Tongé of the City College of New York, and Sharon Nelson of Rutgers.

I would like to thank Marty Levine of MARKET SOURCE for first presenting my books to McGraw-Hill.

I would like to thank McGraw-Hill, especially John Carleo, John Aliano, David Beckwith, and Pat Koch.

I would like to thank Barbara Gilson, Mary Loebig Giles, and Michelle Matozzo Bracci of McGraw-Hill and Marc Campbell of North Market Street Graphics for improving and beautifying the new editions of this series.

I would also like to thank my parents, Lee and Cele, who saw the beginnings of these books but did not live to see their publication.

Last I would like to thank three people who helped keep my spirits up when things looked very bleak: a great friend, Gary Pitkofsky; another terrific friend and fellow lecturer, David Schwinger; and my sharer of dreams, my cousin, Keith Ellis, who also did not live to see my books published.

TO THE STUDENT

This book was written for you: not your teacher, not your next-door neighbor, not for anyone but you. I have tried to make the examples and explanations as clear as I can. However, as much as I hate to admit it, I am not perfect. If you find something that is unclear or should be added to this book, please let me know. If you want a response, or if I can help you, your class, or your school in any precalculus or calculus subject, please let me know, but address your comments c/o McGraw-Hill, Schaum Division, 11 West 19th St., New York, New York 10011.

If you make a suggestion on how to teach one of these topics better and you are the first and I use it, I will give you credit for it in the next edition.

Please be patient on responses. I am hoping the book is so good that millions of you will write. I will answer.

Now, enjoy the book and learn.

CONTENTS

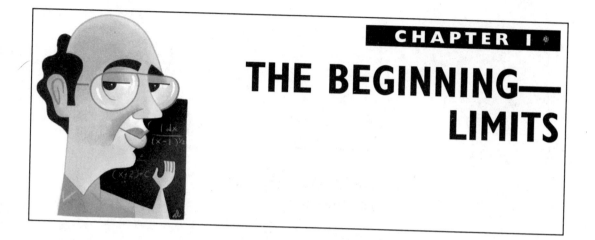

CHAPTER 1

THE BEGINNING— LIMITS

INFORMAL DEFINITION

We will begin at the beginning. Calculus starts with the concept of limits. We will examine this first intuitively before we tackle the more difficult theoretical definition.

Let us examine

$$\lim_{x \to a} f(x) = L;$$

read, "The limit of f(x) as x goes to a is L."

This means that the closer x gets to a, the closer f(x) gets to L. We will leave the word "close" unspecified until later.

EXAMPLE 1—

$$\lim_{x \to 3} 2x$$

We will take points near x = 3, smaller than 3, getting closer to 3. We make a small chart showing this.

x	2x
2.5	5
2.9	5.8
2.99	5.98
2.999	5.998

We see that as x approaches 3 from points less than 3, f(x) approaches 6. *Notation:*

$$\lim_{x \to 3^-} f(x) = 6$$

read, "the limit of f(x) as x goes to 3 from the negative side of 3 (numbers less than 3) is 6." We call this the limit from the left.

   If we do the same thing for numbers greater than 3, the chart would look like this:

x	2x
3.2	6.4
3.1	6.2
3.01	6.02
3.001	6.002

The limit from the right,

$$\lim_{x \to 3^+} f(x),$$

also equals 6. Since the limit from the left equals the limit from the right, the limit exists and is equal to 6. We write

$$\lim_{x \to 3} f(x) = 6.$$

   After seeing this example, you might tell me, "Hey, you big dummy!! All you have to do is substitute x = 3 and get the answer!!" Substitution does work sometimes and should always be tried first. However, if limits (and calculus) were so easy, it would not have taken such dynamite mathematicians as Newton and Leibniz to discover calculus.

EXAMPLE 2—

$$\lim_{x \to 4} \frac{x - 4}{x - 4}$$

We first substitute $x = 4$ and get 0/0, which is indeterminate. We again make a chart.

x	$\dfrac{x - 4}{x - 4}$
4.1	1
4.01	1
4.001	1
3.9	1
3.99	1
3.9999	1

As we get close to 4 from both sides, the answer not only is close to 1 but equals 1. We conclude that the limit as x goes to 4 equals 1.

We get a little better idea of

$$\lim_{x \to a} f(x) = L.$$

This means that $f(x)$ is defined at all points very close to a and that the closer x gets to a, the closer $f(x)$ gets to L (if it doesn't already *equal* L).

EXAMPLE 3—

$$\lim_{x \to 2} \frac{x^2 + 3x + 4}{2x + 5}$$

Nothing bad here.

$$\frac{(2)^2 + 3(2) + 4}{2(2) + 5} = \frac{14}{9}$$

EXAMPLE 4—

$$\lim_{x \to 4} \frac{x - 4}{2x + 3} = \frac{0}{11} = 0$$

EXAMPLE 5—

$$\lim_{x \to 2} \frac{x}{x - 2} = \frac{2}{0},$$

which is undefined.

The limit does not exist. The limit must be a number; infinity is not a number.

Let's give one more demonstrated example of what it is to find the limit point by point.

$$\lim_{x \to 2} \frac{x^2 - 4}{5x - 10}$$

First we let $x = 2$. We find the answer is 0/0. Let's make charts again.

x	$\dfrac{x^2 - 4}{5x - 10}$	x	$\dfrac{x^2 - 4}{5x - 10}$
3	1.0	1	0.6
2.5	0.9	1.5	0.7
2.1	0.82	1.9	0.78
2.01	0.802	1.99	0.798
2.001	0.8002	1.999	0.7998

So

$$\lim_{x \to 2^+} \frac{x^2 - 4}{5x - 10} = 0.8$$

and

$$\lim_{x \to 2^-} \frac{x^2 - 4}{5x - 10} = 0.8$$

Therefore, the limit is 0.8. However, we can't make a chart every time. For Examples 3, 4, and 5, a chart is not necessary. However, Example 6 shows what has to be done sometimes.

Warning: Substitution of a number like $x = 2$ does not work all the time, especially when you have a function that is defined in pieces, such as that in Example 21 at the end of this chapter. Note that $f(1) = 6$, but

$$\lim_{x \to 1} f(x)$$

is 1. Also note that $f(6) = 4$, but the $\lim f(x)$ as x goes to 6 does not exist. So be carrrrreful!!!

EXAMPLE 6—

$$\lim_{x \to 3} \frac{x^2 - 9}{5x - 15}$$

First we substitute $x = 3$ and get 0/0, which is indeterminate. We don't want to make charts all the time. In this case we can factor.

$$\frac{x^2 - 9}{5x - 15} = \frac{(x + 3)(x - 3)}{5(x - 3)}$$

$$\lim_{x \to 3} \frac{(x + 3)}{5} \frac{(x - 3)}{(x - 3)} = \frac{6}{5} \cdot 1 = \frac{6}{5}$$

EXAMPLE 7—

$$\lim_{x \to 0} \frac{x}{|x|}$$

First we substitute x = 0, and we again get 0/0. Making a chart, we get

x	$\dfrac{x}{|x|}$
0.3	1
0.1	1
0.01	1
0.001	1
−0.1	−1
−0.01	−1
−0.0001	−1

The limit from the left is −1, and the limit from the right is 1. Since they are not the same,

$$\lim_{x \to 0} f(x)$$

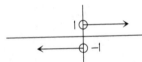

does not exist. The graph will show that the limit does not exist at x = 0.

EXAMPLE 8—

$$\lim_{x \to 4} \frac{x^{1/2} - 2}{x - 4}$$

There are two ways to do this problem. We can rationalize the numerator, remembering not to multiply out the bottom, or we can factor the bottom into the difference of two squares in a kind of weird way not found in most algebra books today.

METHOD A

$$\frac{(x^{1/2} - 2)}{(x - 4)} \frac{(x^{1/2} + 2)}{(x^{1/2} + 2)} = \frac{(x - 4)}{(x - 4)(x^{1/2} + 2)} = \frac{1}{x^{1/2} + 2}$$

If we now take

$$\lim_{x \to 4} \frac{1}{(x^{1/2} + 2)} = \frac{1}{(2 + 2)} = 1/4$$

METHOD B

$$\frac{x^{1/2} - 2}{x - 4} = \frac{x^{1/2} - 2}{(x^{1/2} - 2)(x^{1/2} + 2)} = \frac{1}{x^{1/2} + 2}$$

which gives the same result.

EXAMPLE 9—

$$\lim_{x \to 3} \frac{(5 - 15/x)}{x - 3}$$

We will multiply top and bottom by x.

$$\frac{(5 - 15/x)x}{(x - 3)x} = \frac{5x - 15}{(x - 3)x} = \frac{5(x - 3)}{x(x - 3)} = \frac{5}{x}$$

Taking

$$\lim_{x \to 3} \frac{5}{x} = \frac{5}{3}$$

LIMITS AS x GOES TO INFINITY

Although this topic occurs later in your book (and my book), some texts talk about limits at infinity very early on. So I've decided to add this section. If you don't need it now, skip it until later.

We need to know one fact from elementary algebra. The *degree* of a polynomial in one unknown is the highest exponent.

EXAMPLE 10—

$$\lim_{x \to \infty} \frac{4x + x^2}{5x^3 + 1}$$

Divide every term, top and bottom, by x^3, x to the higher degree.

$$\frac{4x + x^2}{5x^3 + 1} = \frac{4x/x^3 + x^2/x^3}{5x^3/x^3 + 1/x^3} = \frac{4/x^2 + 1/x}{5 + 1/x^3}$$

If we now take the limit as x goes to infinity, every term goes to 0 except the 5 and we get $(0 + 0)/(5 + 0) = 0$.

IMPORTANT NOTE I

Anytime the degree of the top is less than the degree of the bottom, the limit will be 0. You need not do the work (unless the teacher demands it). You should know that the limit is 0!!!!!!

EXAMPLE I I

$$\lim_{x \to \infty} \frac{x^2 x^{1/2} + 3x - 1}{x^3 + 5x^2 - 99}$$

If we could talk about a degree of the top, it would be 5/2, or 2½. Since the degree of the bottom is 3, which is more than the top, the limit is 0!

EXAMPLE I 2

Divide everything by x^3.

$$\lim_{x \to \infty} \frac{4x^3 + 7}{x - 3x^3}$$

We get

$$\frac{4x^3/x^3 + 7/x^3}{x/x^3 - 3x^3/x^3} = \frac{4 + 7/x^3}{1/x^2 - 3}$$

If we now let x go to infinity, we get the limit to be 4/−3 or −4/3.

IMPORTANT NOTE 2

If the degree of the top is the same as the degree of the bottom, the limit is the coefficient of the highest power on the top divided by the coefficient of the highest power on the bottom. Again, you do not actually have to do the division.

Here are two more limits as x goes to infinity.

EXAMPLE I 3

$$\lim_{x \to \infty} (x^2 + 4)^{1/2} - (x^2 - 1)^{1/2}$$

We get infinity minus infinity. No good!!! What to do? We rationalize the situation. Seriously, we multiply top and bottom by the conjugate.

$$[(x^2 + 4)^{1/2} - (x^2 - 1)^{1/2}] \frac{(x^2 + 4)^{1/2} + (x^2 - 1)^{1/2}}{(x^2 + 4)^{1/2} + (x^2 - 1)^{1/2}}$$

So we get

$$\lim_{x \to \infty} \frac{5}{(x^2 + 4)^{1/2} + (x^2 - 1)^{1/2}} = 0$$

EXAMPLE 14, PART A—

$$\lim_{x \to \infty} \frac{(3x^2 + 4x + 5)^{1/2}}{7x}$$

EXAMPLE 14, PART B—

Same example, except x goes to minus infinity.

As x goes to plus or minus infinity, only the highest power of x counts. So $(3x^2 + 4 + 5)^{1/2}$ is approximately equal to $3^{1/2} |x|$ for very big and very small values of x.

A. Soooo

$$\lim_{x \to \infty} \frac{3^{1/2} |x|}{7x} = \frac{3^{1/2}}{7}$$

buuuuuuut

B. $$\lim_{x \to -\infty} \frac{3^{1/2} |x|}{7x} = \frac{-3^{1/2}}{7}!!$$

PROBLEMS INVOLVING $\lim_{x \to 0} \dfrac{\sin x}{x}$

In proving that the derivative of the sine is the cosine, which is found in nearly every text, we also prove

$$\lim_{x \to 0} (\sin x)/x = 1.$$

This means if we take the sine of any angle and divide it by *precisely* the same angle, if we now take the limit as we go to 0, the value is 1. For some reason, this topic, which requires almost no writing or calculation, causes a tremendous amount of agony. Hopefully I can lessen the pain.

FACT I

$\lim_{x \to 0} \sin x/x = 1$

FACT 2

$\lim_{x \to 0} \sin x = 0$

FACT 3

$\lim_{x \to 0} \cos x = 1$

EXAMPLE 15

To use Fact 1, since the angle on the top is 3x, the angle on the bottom must also be 3x. So put the 4 on the left and multiply the bottom by the 3 you need. If you multiply the bottom by 3, you must multiply the top by 3 so nothing changes.

$\lim_{x \to 0} \frac{\sin 3x}{4x}$

Sooooo . . .

$$\frac{\sin 3x}{4x} = \frac{3}{3}\frac{\sin 3x}{4x} = \frac{3}{4}\frac{\sin 3x}{3x}$$

$$\lim_{x \to 0} \frac{3}{4}\frac{\sin 3x}{3x} = \frac{3}{4}\lim_{x \to 0}\frac{\sin 3x}{3x} = \frac{3}{4}(1) = \frac{3}{4}$$

We use the identity tan x = sin x/cos x.

EXAMPLE 16

$\lim_{x \to 0} \tan x/x$

$$\frac{\tan x}{x} = \frac{1}{x}\frac{\sin x}{\cos x} = \frac{1}{\cos x}\frac{\sin x}{x}$$

Therefore

$$\lim_{x\to 0}\frac{\tan x}{x} = \lim_{x\to 0}\frac{1}{\cos x}\frac{\sin x}{x} = \left(\lim_{x\to 0}\frac{1}{\cos x}\right)\left(\lim_{x\to 0}\frac{\sin x}{x}\right)$$

$$= (1/1) \times 1 = 1.$$

How are you doing so far? Let's put in one more example.

EXAMPLE 17—

$$\lim_{x\to 0}\frac{\sin^2 x}{x} = \lim_{x\to 0}\sin x \frac{\sin x}{x} = (\lim_{x\to 0}\sin x)\left(\lim_{x\to 0}\frac{\sin x}{x}\right)$$

$$= 0(1) = 0$$

NOTE

In Example 3, if $\sin^2 x$ were in the bottom and x were in the top, then the limit would be 1/0, which would be undefined.

FORMAL DEFINITION

We will now tackle the most difficult part of basic calculus, the theoretical definition of limit. As previously mentioned, it took two of the finest mathematicians of all times, Newton and Leibniz, to first formalize this topic. It is *not* essential to the rest of basic calculus to understand this definition. It is hoped this explanation will give you some understanding of how really amazing calculus is and how brilliant Newton and Leibniz must have been. Remember this is an approximating process that many times gives exact (or if not, very, very close) answers. To me this is mind-boggling, terrific, stupendous, unbelievable, awesome, cool, and every other great word you can think of.

DEFINITION

$$\lim_{x \to a} f(x) = L$$

if and only if, given $\varepsilon > 0$, there exists $\delta > 0$ such that if $0 < |x - a| < \delta$, then $|f(x) - L| < \varepsilon$.

NOTE

ε = epsilon and δ = delta—two letters of the Greek alphabet.

TRANSLATION I

Given ε, a small positive number, we can always find δ, another small positive number, such that if x is within a distance δ from a but not exactly at a, then f(x) is within a distance ε from L.

TRANSLATION 2

We will explain this definition using an incorrect picture. I feel this gives you a much better idea than the correct picture, which we will use next.

Interpret $|x - a|$ as the distance between x and a, but instead of the one-dimensional picture it really is, imagine that there is a circle around the point a of radius δ. $|x - a| < \delta$ stands for all x values that are inside this circle. Similarly, imagine a circle of radius ε around L, with $|f(x) - L| < \varepsilon$ the set of all points f(x) that are inside this circle.

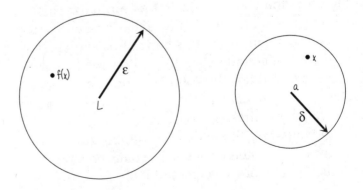

The definition says given $\varepsilon > 0$ (given a circle of radius ε around L), we can find $\delta > 0$ (circle of radius δ around a) such that if $0 < |x - a| < \delta$ (if we take any x inside this circle), then $|f(x) - L| < \varepsilon$ (f(x)) will be inside of the circle of radius ε but not exactly at L.

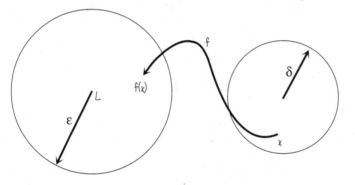

Now take another ε ε_2, positive but smaller than ε (a smaller circle around L); there exists another δ δ_2, usually a smaller circle around a, such that if $0 < |x - a| < \delta_2$, then $|f(x) - L| < \varepsilon_2$.

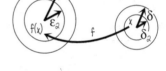

Now take smaller and smaller positive ε's; we can find smaller and smaller δ's. In the limit as the x circle shrinks to a, the f(x) circle shrinks to L. *Read this a number of times!!!*

TRANSLATION 3

Let us see the real picture. $y = f(x)$. $|x - a| < \delta$ means $a - \delta < x < a + \delta$. $|y - L| < \varepsilon$ means $L - \varepsilon < y < L + \varepsilon$.

Given $\varepsilon > 0$, if we take any x value such that $0 < |x - a| < \delta$, the interval on the x axis, and find the corresponding $y = f(x)$ value, this y value must be within ε of L, that is, $|f(x) - L| < \varepsilon$.

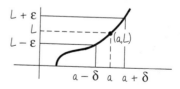

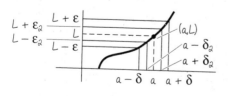

Take a smaller ε_2; we can find δ_2 such that $0 < |x - a| < \delta_2$, $|f(x) - L| < \varepsilon_2$. The smaller the ε, the smaller the δ. $f(x)$ goes to L as x goes to a.

Although this definition is extremely difficult, its application is pretty easy. We need to review six facts, four about absolute value and two about fractions.

1. $|ab| = |a|\,|b|$.

2. $\left|\dfrac{a}{b}\right| = \dfrac{|a|}{|b|}$.

3. $|a - b| = |b - a|$.

4. $|a + b| \leq |a| + |b|$.

5. In comparing two positive fractions, if the bottoms are the same and both numerators and denominators are positive, the larger the top, the larger the fraction. $2/7 < 3/7$.

6. If the tops are the same, the larger the bottom, the smaller the fraction. $3/10 > 3/11$.

Now let's do some problems.

EXAMPLE 18—

Using ε, δ, prove

$$\lim_{x \to 2} (4x - 3) = 5.$$

In the definition

$$\lim_{x \to a} f(x) = L$$

$f(x) = 4x - 3$, $a = 2$, $L = 5$. Given $\varepsilon > 0$, we must find $\delta > 0$, such that if $0 < |x - 2| < \delta$, $|(4x - 3) - 5| < \varepsilon$.

$$|(4x - 3) - 5| = |4x - 8| = |4(x - 2)|$$
$$= |4|\,|x - 2| < 4 \cdot \delta = \varepsilon \qquad \delta = \varepsilon/4$$

EXAMPLE 19—

Prove

$$\lim_{x \to 4} (x^2 + 2x) = 24.$$

Given $\varepsilon > 0$, we must find $\delta > 0$, such that if $0 < |x - 4| < \delta$, $|x^2 + 2x - 24| < \varepsilon$.

$|x^2 + 2x - 24| = |(x + 6)(x - 4)| = |x + 6|\ |x - 4|$

We must make sure that $|x + 6|$ does not get too big. We must always find δ, no matter how small. We must take a preliminary $\delta = 1$. $|x - 4| < 1$, which means $-1 < x - 4 < 1$ or $3 < x < 5$. In any case $x < 5$. Sooooo . . .

$|x + 6| \le |x| + |6| < 5 + 6 = 11$

Finishing our problem, $|x + 6|\ |x - 4| < 11 \cdot \delta = \varepsilon$. $\delta =$ minimum $(1, \varepsilon/11)$.

EXAMPLE 20—

Prove

$$\lim_{x \to 5} \frac{2}{x} = \frac{2}{5}.$$

$$\left|\frac{2}{x} - \frac{2}{5}\right| = \left|\frac{10 - 2x}{5x}\right| = \frac{|2(5 - x)|}{|5x|} = \frac{|2|\ |5 - x|}{|5|\ |x|}$$

$$= \frac{2|x - 5|}{5|x|}$$

Again take a preliminary $\delta = 1$. $|x - 5| < 1$. So $4 < x < 6$. To make a fraction larger, make the top larger and the bottom smaller. $0 < |x - 5| < \delta$. We substitute δ on the top. Since $x > 4$, we substitute 4 on the bottom.

$$\frac{2|x - 5|}{5|x|} < \frac{2 \cdot \delta}{5 \cdot 4} = \frac{\delta}{10} = \varepsilon \qquad \delta = 10\varepsilon$$

If $\delta =$ minimum $(1, 10\varepsilon)$, $|2/x - 2/5| < \varepsilon$.

CONTINUITY

We finish with a brief discussion of continuity of a function at a point. Intuitively, continuity at a point means there is no break in the graph of the function at the point. Let us define continuity more formally:

f(x) is *continuous* at point a if

1. $\lim\limits_{x \to a} f(x) = L$

2. $f(a) = L$

We will do a longish example to illustrate the definition fully.

EXAMPLE 21—

Let f(x) = 1   x < 1

= 6   x = 1

= x   1 < x < 3

= 6 − x   3 ≤ x < 6

= 4   x ≥ 6

We wish to examine the continuity at x = 1, 3, and 6. Let's graph this function.

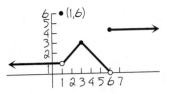

At x = 1, the limit from the left of f(x) = 1 is 1. The limit from the right of f(x) = x is 1. So

$\lim\limits_{x \to 1} f(x)$

exists and equals 1. Part 1 of the definition is satisfied. But f(1) = 6. The function is not continuous at 1. (See the jump to y = 6 at x = 1.)

At x = 3, the limits from the left and the right at 3 equal 3. In addition, f(3) = 3. The function is continuous at x = 3. (Notice, no break at x = 3.)

At $x = 6$, the limit as x goes to 6 from the left is 0. The limit as x goes to 6 from the right is 4. Since the two are different, the limit does not exist. The function is not continuous at 6 (see the jump). We do not have to test the second part of the definition since part 1 fails.

CHAPTER 2

THE BASICS

DERIVATIVES—DEFINITION AND RULES

We would like to study the word *tangent.* In the case of a circle, the line L_1 is tangent to the circle if it hits the circle in one and only one place.

In the case of a general curve, we must be more careful. We wish to exclude lines like L_2. We wish to include lines like L_3, even though, if extended, such a line would hit the curve again.

We also need to use the word *secant.* L_4 is secant to a circle if it hits the curve in two places.

DEFINITION

Tangent line to a curve at the point P.

A. Take point P on the curve.

B. Take point Q_1 on the curve.

C. Draw PQ_1.

D. Take $Q_2, Q_3, Q_4, \ldots$, drawing $PQ_2, PQ_3, PQ_4, \ldots$ with Q's approaching P.

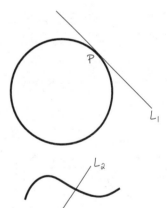

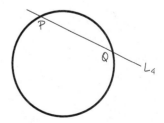

E. Do the same thing on the other side of P: R_1, R_2, . . . such that R_1 and R_2 are approaching P.

F. If the secant lines on each side approach one line, we will say that this line is tangent to the curve at point P.

We would like to develop the idea of tangent algebraically. We will review the development of slope from algebra.

Given points P_1—coordinates (x_1,y_1)—and P_2—coordinates (x_2,y_2). Draw the line segment through P_1 parallel to the x axis and the line segment through P_2 parallel to the y axis, meeting at point Q. Since everything on P_1Q has same y value, the y coordinate of Q is y_1.

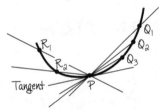

Everything on P_2Q has the same x value. The x value of Q is x_2. The coordinates of Q are (x_2,y_1).

Since everything on P_1Q has the same y value, the length of $P_1Q = x_2 - x_1$. Since everything on P_2Q has the same x value, the length of $P_2Q = y_2 - y_1$. The slope

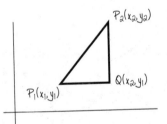

$$m = \frac{\text{change in y}}{\text{change in x}} = \frac{y_2 - y_1}{x_2 - x_1} = \frac{\Delta y}{\Delta x}$$

Δ = delta, another Greek letter.

Let's do the same thing for a general function $y = f(x)$.

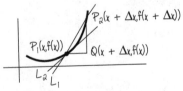

Let point P_1 be the point $(x,y) = (x,f(x))$. A little bit away from x is $x + \Delta x$. (We drew it a lot away; otherwise you could not see it.) The corresponding y value is $f(x + \Delta x)$. So $P_2 = (x + \Delta x, f(x + \Delta x))$. As before, draw a line through P_1 parallel to the x axis and a line through P_2 parallel to the y axis. The lines again meet at Q. As before, Q has the same x value as P_2 and the same y value as P_1. Its coordinates are $(x + \Delta x, f(x))$. Since all y values on P_1Q are the same, the length of $P_1Q = (x + \Delta x) - x = \Delta x$. All x values on P_2Q are the same. The length of $P_2Q = f(x + \Delta x) - f(x)$. The slope of the secant line

$$L_1 = \frac{\Delta y}{\Delta x} = \frac{P_2Q}{P_1Q} = \frac{f(x + \Delta x) - f(x)}{\Delta x}$$

Now we do as before—let P_2 go to P_1. Algebraically this means to take the limit as Δx goes to 0. We get the slope of the tangent line L_2 at P_1. Our notation will be . . .

The slope of the tangent line

$$L_2 = \lim_{\Delta x \to 0} \frac{f(x + \Delta x) - f(x)}{\Delta x}$$

if it exists.

DEFINITION

Suppose $y = f(x)$. The derivative of $f(x)$, at a point x denoted b, $f'(x)$, or dy/dx is defined as

$$\lim_{\Delta x \to 0} \frac{f(x + \Delta x) - f(x)}{\Delta x}$$

if it exists.

NOTE 1

All mathematics originally came from a picture. The idea of derivative came from the slope. Now the definition is independent of the picture.

NOTE 2

If $y = f(t)$ is a distance as a function of time t

$$\lim_{\Delta t \to 0} \frac{f(t + \Delta t) - f(t)}{\Delta t} = f'(t)$$

is the velocity y(t).

Well, heck. Note 2 about velocity is not enough! Let's do some examples.

EXAMPLE 1—

Suppose y = f(t) stands for the distance at some point in time t. Then f(t + Δt) stands for your location later, if Δt is positive. (Remember, Δt means a change in time.) y = f(t + Δt) − f(t) is the distance traveled in time Δt.

$$\frac{\Delta y}{\Delta t} = \frac{f(t + \Delta t) - f(t)}{\Delta t} = \text{average velocity}$$

If we take the limit as Δt goes to 0, that is,

$$\lim_{\Delta t \to 0} \frac{f(t + \Delta t) - f(t)}{\Delta t},$$

then f'(t) is the instantaneous velocity at any time t.

NOTE 1

The average velocity is very similar to the rate you learned in elementary algebra. If you took the distance traveled and divided it by the time, you would get the rate. The only difference is that in algebra, the average velocity was always the same.

NOTE 2

Even if you drive a car at 30 mph, at any instant you might be going a little faster or slower. This is the instantaneous velocity.

EXAMPLE 2—

Let $f(t) = t^2 + 5t$, f(t) in feet, t in seconds.

A. Find the distance traveled between the third and fifth seconds.

B. Find the average velocity $3 \le t \le 5$.

C. Find the instantaneous velocity at t = 5.

A. $\Delta y = f(t + \Delta t) - f(t) = (t + \Delta t)^2 + 5(t + \Delta t) - (t^2 + 5t)$

$t = 3; \Delta t = 5 - 3 = 2$

$= t^2 + 2t \Delta t + (\Delta t)^2 + 5t + 5 \Delta t - t^2 - 5t$

$= 2t \Delta t + (\Delta t)^2 + 5 \Delta t = 2(3)(2) + (2)^2 + 5(2)$

$= 26$ feet

B. $v_{av} = \Delta y / \Delta t = 26/2 = 13$ feet per second

C. $v_{inst} = \lim\limits_{\Delta t \to 0} \dfrac{f(t + \Delta t) - f(t)}{\Delta t} = \lim\limits_{\Delta t \to 0} \dfrac{2t \Delta t + (\Delta t)^2 + 5 \Delta t}{\Delta t}$

$= \lim\limits_{\Delta t \to 0} (2t + \Delta t + 5) = 2t + 5$

At $t = 5$, $v_{inst} = 2t + 5 = 2(5) + 5 = 15$ feet per second.

NOTE I

Δt is usually very small when compared to t.

NOTE 2

The derivative does not always exist. If $y = |x|$, the derivative does not exist at $x = 0$, since all secant lines on the left have slope −1 and all on the right have slope 1. These lines never approach one line.

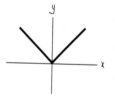

In almost all courses, you are asked to do some problems using the definition of *derivative*. This is really a thorough exercise in algebra with just a touch of limits. Let us do three examples.

EXAMPLE 3—

Using the definition of *derivative*, find f′(x) if f(x) = $3x^2 + 4x - 5$.

$f(x + \Delta x) = 3(x + \Delta x)^2 + 4(x + \Delta x) - 5$

$f(x + \Delta x) - f(x) = 3(x + \Delta x)^2 + 4(x + \Delta x) - 5 - (3x^2 + 4x - 5)$

$= 3x^2 + 6x \Delta x + 3(\Delta x)^2 + 4x + 4 \Delta x - 5$

$- 3x^2 - 4x + 5$

$= 6x \Delta x + 3(\Delta x)^2 + 4 \Delta x$

If you have done your algebra correctly, all remaining terms at this point will have at least Δx multiplying them. If there is a fraction, all terms in the top will have at least Δx.

$$\frac{f(x + \Delta x) - f(x)}{\Delta x} = \frac{6x\,\Delta x + 3(\Delta x)^2 + 4\,\Delta x}{\Delta x} = 6x + 3\,\Delta x + 4$$

$$f'(x) = \lim_{\Delta x \to 0} (6x + 3\,\Delta x + 4) = 6x + 4$$

If $f(x) = 3x^2 + 4x - 5$ were a curve, the slope of the tangent line at any point on the curve would be found by multiplying the x value by 6 and adding 4.

EXAMPLE 3 CONTINUED—

We don't need to use the 34.

Find the slope of the tangent line to the curve $f(x) = 3x^2 + 4x - 5$ at the point (3,34).

The slope $m = 6x + 4 = 6(3) + 4 = 22$.

EXAMPLE 3 CONTINUED, CONTINUED—

Find the equation of the line tangent to $f(x) = 3x^2 + 4x - 5$ at the point (3,34).

From algebra, the equation of a line is given by

$$m = \frac{y - y_1}{x - x_1}$$

(point-slope). $x_1 = 3$, $y_1 = 34$, and the slope m is $f'(3) = 22$. The equation of the line is

$$22 = \frac{y - 34}{x - 3},$$

which you can simplify if forced to.

EXAMPLE 3 LAST CONTINUATION—

Find the equation of the line normal to $y = 3x^2 + 4x - 5$ at the point (3,34).

The word *normal* means draw the tangent line at point P and then draw the perpendicular to that tangent line at point P. Perpendicular slope means negative reciprocal. The equation of the normal line is . . .

$$\frac{-1}{22} = \frac{y - 34}{x - 3}$$

EXAMPLE 4—

$$f(x) = \frac{x}{x + 5}$$

Find f'(x) using the definition of *derivative.*

$$\frac{f(x + \Delta x) - f(x)}{\Delta x} = \frac{\dfrac{x + \Delta x}{x + \Delta x + 5} - \dfrac{x}{x + 5}}{\Delta x}$$

$$= \frac{(x + \Delta x)(x + 5) - x(x + \Delta x + 5)}{(x + \Delta x + 5)(x + 5)(\Delta x)}$$

$$= \frac{5\,\Delta x}{(x + \Delta x + 5)(x + 5)(\Delta x)}$$

$$= \frac{5}{(x + \Delta x + 5)(x + 5)}$$

$$f'(x) = \lim_{\Delta x \to 0} \frac{5}{(x + \Delta x + 5)(x + 5)} = \frac{5}{(x + 5)(x + 5)} = \frac{5}{(x + 5)^2}$$

We must multiply out the top, but we do not multiply the bottom.

EXAMPLE 5—

Find g'(x) using the definition of *derivative* if

$$g(x) = \sqrt{x}.$$

$$\frac{g(x + \Delta x) - g(x)}{\Delta x} = \frac{\sqrt{x + \Delta x} - \sqrt{x}}{\Delta x}$$

Rationalize the numerator.

$$= \frac{(\sqrt{x + \Delta x} - \sqrt{x})}{\Delta x} \cdot \frac{(\sqrt{x + \Delta x} + \sqrt{x})}{(\sqrt{x + \Delta x} + \sqrt{x})}$$

$$= \frac{(x + \Delta x) - x}{(\Delta x)(\sqrt{x + \Delta x} + \sqrt{x})}$$

$$= \frac{\Delta x}{(\Delta x)(\sqrt{x + \Delta x} + \sqrt{x})}$$

$$= \frac{1}{\sqrt{x + \Delta x} + \sqrt{x}}$$

$$g'(x) = \lim_{\Delta x \to 0} \frac{1}{\sqrt{x + \Delta x} + \sqrt{x}}$$

$$= \frac{1}{\sqrt{x} + \sqrt{x}} = \frac{1}{2\sqrt{x}}$$

We can't keep using the definition of *derivative*. If we had a complicated function, it would take forever. We will list the rules, interpret them, and give examples. Proofs are found in most calculus books.

RULE 1

If $f(x) = c$, $f'(x) = 0$. The derivative of a constant is 0.

RULE 2

If $f(x) = x$, $f'(x) = 1$.

RULE 3

If $f(x) = x^n$, $f'(x) = nx^{n-1}$. Bring down the exponent and subtract 1 to get the new exponent.

RULE 4

If $f(x) = cg(x)$, $f'(x) = cg'(x)$. If we have a constant c multiplying a function, we leave c alone and only take the derivative of the function.

RULE 5

If $f(x) = g(x) \pm h(x)$, then $f'(x) = g'(x) \pm h'(x)$.

In a book, t, u, v, w, x, y, and z are usually variables. a, b, c, and k (since mathematicians can't spell) are used as constants.

EXAMPLE 6—

$y = 3x^7 + 7x^4 + 2x + 3$. Find y'.

$y' = 21x^6 + 28x^3 + 2 \ (+0)$

EXAMPLE 7—

Find y' if

$$y = 3x^4 + \frac{a}{x^6} + \frac{1}{bx^7} + \sqrt[4]{x^9} + \frac{1}{\sqrt[11]{x^5}} + \pi^7$$

$$y = 3x^4 + ax^{-6} + \frac{1}{b}x^{-7} + x^{9/4} + x^{-5/11} + \pi^7$$

$$y' = 12x^3 - 6ax^{-7} + \frac{1}{b}(-7x^{-8}) + (9/4)x^{5/4} + (-5/11)x^{-16/11} + 0$$

$$y' = 12x^3 - \frac{6a}{x^7} - \frac{7}{bx^8} + \frac{9x^{5/4}}{4} - \frac{5}{11x^{16/11}}$$

a and l/b are constants.
9/4 − 1 = 5/4.

−5/11 − 1 = −16/11. Derivative of messy constants is still 0.

Most calculus books give the derivative of the six trigonometric functions near the beginning. So will we.

RULE 6

    A. If $y = \sin x$, $y' = \cos x$.

    B. If $y = \cos x$, $y' = -\sin x$.

    C. If $y = \tan x$, $y' = \sec^2 x$.

    D. If $y = \cot x$, $y' = -\csc^2 x$.

    E. If $y = \sec x$, $y' = \tan x \sec x$.

    F. If $y = \csc x$, $y' = -\cot x \csc x$.

RULE 7

The product rule. If $y = f(x)g(x)$, then $y' = f(x)g(x)' + g(x)f(x)'$. The product rule says the first multiplied by the derivative of the second added to the second multiplied by the derivative of the first.

EXAMPLE 8—

Find y' if $y = (x^2 + 3x + 1)(5x + 2)$.

$$\qquad\qquad\text{(first)}\quad\text{(second)}$$

$$y' = (x^2 + 3x + 1)\quad (5)\quad + (5x + 2)(2x + 3)$$

$$\quad\text{(first)}\quad\text{(second)}' + \text{(second)}\ \text{(first)}'$$

Multiplying and combining like terms, $y' = 15x^2 + 34x + 11$.

We could, of course multiply this example out.

$y = 5x^3 + 17x^2 + 11x + 2$.

Then $y' = 15x^2 + 34x + 11$ as before.

However, later on the examples will be much longer or even impossible to multiply out. It is a blessing that we have the product rule and the next two rules.

RULE 8

The quotient rule.

If $y = f(x)/g(x)$, then $y' = \dfrac{g(x)f'(x) - f(x)g'(x)}{[g(x)]^2}$

The quotient rule says the bottom times the derivative of the top minus the top times the derivative of the bottom, all divided by the bottom squared.

EXAMPLE 9—

Find y' if $y = \dfrac{x^2}{x^3 + 2x + 1}$.

(bottom)(top)′
− (top)(bottom)′
────────────
bottom squared

$$y' = \frac{(x^3 + 2x + 1)(2x) - x^2(3x^2 + 2)}{(x^3 + 2x + 1)^2}$$

When simplifying, do *not* multiply out the bottom. Only multiply and simplify the top. You may simplify the top by factoring, as we will do in other problems.

Simplified, $y' = \dfrac{-x^4 + 2x^2 + 2x}{(x^3 + 2x + 1)^2}$.

RULE 9

The chain rule. Suppose we have a composite function $y = f(u)$, $u = u(x)$.

Then $\dfrac{dy}{dx} = \dfrac{dy}{du} \cdot \dfrac{du}{dx}$.

EXAMPLE 10—

Let $f(x) = (x^2 + 1)^{100}$.

One way is to multiply this out. We dismiss this on grounds of sanity.

We let $u = x^2 + 1$. Then $y = f(u) = u^{100}$.

$$\frac{du}{dx} = 2x. \qquad \frac{dy}{du} = 100u^{99}$$

Then $\dfrac{dy}{dx} = \dfrac{dy}{du}\dfrac{du}{dx} = (100u^{99})(2x) = 100(x^2 + 1)^{99} \cdot 2x$

$$= 200x(x^2 + 1)^{99}.$$

We don't want to write u each time. We will imagine what u is and use the chain rule. Try it. It only takes a little practice.

EXAMPLE 11—

Find y' if $y = (x^3 + 7x^2 + 1)^{4/3}$.

Imagine $u = x^3 + 7x^2 + 1$. $\dfrac{du}{dx} = 3x^2 + 14x$.

$y' = (4/3)(x^3 + 7x^2 + 1)^{1/3}(3x^2 + 14x)$.

EXAMPLE 12—

$y = \tan (x^4 + 3x - 11)$.

Imagine $u = x^4 + 3x - 11$. $y' = \sec^2 (x^4 + 3x - 11) \cdot (4x^3 + 3)$.

EXAMPLE 13—

Find y' if $y = \sin^6 (x^4 + 3x)$.

This is a double composite: a function of a function of a function. We use the chain rule twice.

Let the crazy angle = $v = x^4 + 3x$. So $dv/dx = 4x^3 + 3$.

Let $u = \sin(x^4 + 3x) = \sin v$. So $du/dv = \cos v$.

So $y = u^6$ and $dy/du = 6u^5$. So . . .

$dy/dx = (dy/du)$ times (du/dv) times (dv/dx)

$= 6u^5$ times $\cos v$ times $4x^3 + 3$

$= [6 \sin^5(x^4 + 3x)][\cos(x^4 + 3x)](4x^3 + 3)$

power rule—	derivative	derivative
leave trig	of trig	of the
function and	function—	crazy angle
crazy angle	crazy angle	
alone	stays	

NOTE

This is *not* the product rule.

Sometimes you will have other combinations of the rules. After a short while, you will find the rules relatively easy. However, the algebra does require practice.

EXAMPLE 14

Find y' if $y = (x^2 + 1)^8 (6x + 7)^5$.

This problem involves the product rule. But in each derivative, we will have to use the chain rule.

$$y' = \underbrace{\frac{(x^2 + 1)^8 \cdot 5(6x + 7)^4(6)}{(\text{first}) \cdot (\text{second})'}} + \underbrace{\frac{(6x + 7)^5 \cdot 8(x^2 + 1)^7(2x)}{(\text{second}) \cdot (\text{first})'}}$$

The calculus is now finished. We now must simplify by factoring. There are two terms, each underlined. From each we must take out the largest common factor. The largest number that can be factored out is 2. No x can be factored out. The lowest powers of $x^2 + 1$ and $6x + 7$ can be factored out. We take out $(x^2 + 1)^7$ and $(6x + 7)^4$.

$$y' = 2(x^2 + 1)^7(6x + 7)^4[15(x^2 + 1) + 8x(6x + 7)]$$

$$= 2(x^2 + 1)^7(6x + 7)^4(63x^2 + 56x + 15)$$

Let us try one more using the quotient rule and chain rule.

EXAMPLE 15

Find y' if $y = \dfrac{x^3}{(x^2 + 1)^4}$.

$$y' = \frac{(x^2 + 1)^4 3x^2 - x^3[4(x^2 + 1)^3(2x)]}{[(x^2 + 1)^4]^2}$$

$$= \frac{x^2(x^2 + 1)^3[3(x^2 + 1) - 8x^2]}{(x^2 + 1)^8}$$

$$= \frac{x^2(-5x^2 + 3)}{(x^2 + 1)^5}$$

When *not* to use the product or quotient rule:

EXAMPLE A

$$y = 5(x^2 - 4)^{10}$$

$$y' = 5 \quad [10(x^2 - 4)^9(2x)] = 100x(x^2 - 4)^9$$

Do not use product rule since 5 is a constant. Only the chain rule is necessary.

EXAMPLE B

$$y = 7/x^5$$

$$y' = 7(-5)x^{-6} = -35/x^6$$

Do not use quotient rule. Rewrite example as $y = 7x^{-5}$ and simply use power rule.

EXAMPLE C

$$y = b/(x^2 + 5)^8$$

$$y' = b(-8)(x^2 + 5)^{-9}(2x) = -16bx/(x^2 + 5)^9$$

Rewrite as $y = b(x^2 + 5)^{-8}$.

IMPLICIT DIFFERENTIATION

Suppose we are given $y^3 + x^4y^7 + x^3 = 9$. It would be difficult, maybe impossible, to solve for y. However, there is a theorem called the implicit function theorem that gives conditions that will show that y = f(x) exists even if we can never find y. Moreover, it will allow us to find dy/dx even if we can never find y. Pretty amazing, isn't it?!!!!!

Let f(y) = y^n where y = y(x). Using the chain rule and power rule, we get

$$\frac{df}{dx} = \frac{df(y)}{dy}\frac{dy(x)}{dx} = ny^{n-1}\frac{dy}{dx}$$

EXAMPLE 16—

Find dy/dx if $y^3 + x^4y^7 + x^3 = 9$.

1. We will differentiate straight across implicitly.

2. We will differentiate the first term implicitly, the second term implicitly and with the product rule, and the rest the old way.

3. We will solve for dy/dx using an algebraic trick that can save up to five algebraic steps. With a little practice, you can save a lot of time!!!!

Let's do the problem.

$$3y^2\frac{dy}{dx} + \left[x^4(7y^6)\frac{dy}{dx} + y^7(4x^3)\right] + 3x^2 = 0$$

| implicit (n = 3) | implicity (n = 7) and product rule | power rule | derivative of constant |

We now solve for dy/dx. Once we take the derivative, it becomes an elementary algebra equation in which we solve for dy/dx.

1. All the terms without dy/dx go to the other (right) side and change signs.

2. All terms without dy/dx on the other side stay there; no sign change.

3. All terms with dy/dx on the right go to the left and change signs.

4. All terms with dy/dx on the left stay there; no sign change.

5. Factor out dy/dx from all terms on the left; this coefficient is divided on both sides. Therefore it goes to the bottom of the fraction on the right.

6. Rearrange all terms so that the number is first and each letter occurs alphabetically.

It really is easy with a little practice. Using this method, our answer is

$$\frac{dy}{dx} = \frac{-4x^3y^7 - 3x^2}{3y^2 + 7x^4y^6}$$

EXAMPLE 16 CONTINUED—

Maybe you think this is too complicated. Let's do another with much simpler coefficients after taking the derivative: $Ay' + B - Cy' - D = 0$.

B and D have no y' and must go to the other side and flip signs. The $(A - C)$ is factored out from the y' and goes to the bottom of the answer. So $y' = (D - B)/(A - C)$.

Still don't believe? Let's do it step by step.

$$\begin{array}{rcl} Ay' + B - Cy' - D &=& 0 \\ \underline{\quad - B \qquad\quad + D} &=& \underline{D - B} \\ Ay' \quad - Cy' \quad &=& D - B \end{array}$$

$$\frac{(A - C)y'}{A - C} = \frac{D - B}{A - C}$$

So

$$y' = \frac{D - B}{A - C}$$

If you look closely, this really is Example 14, the algebraic part.

EXAMPLE 16 LAST CONTINUATION

Suppose we are given the same equation, but are asked to find dy/dx at the point (−2,1).

The first step is the same.

$$3y^2 \frac{dy}{dx} + x^4\left(7y^6 \frac{dy}{dx}\right) + y^7(4x^3) + 3x^2 = 0$$

But instead of doing all the rest of the work, we substitute x = −2 and y = 1!

$$3 \frac{dy}{dx} + 112 \frac{dy}{dx} - 32 + 12 = 0$$

So dy/dx = 20/115 = 4/23. We do as little work as possible. Mathematicians are very lazy by nature.

Notations

We know the notations for the first derivative. Suppose we want to take more derivatives . . .

$$y' \quad = f'(x) \quad = \frac{dy}{dx} \quad = \text{first derivative}$$

$$y'' \quad = f''(x) \quad = \frac{d^2y}{dx^2} \quad = \text{second derivative}$$

$$y^{(27)} \quad = f^{(27)}(x) \quad = \frac{d^{27}y}{dx^{27}} \quad = \text{twenty-seventh derivative}$$

$$y^{(n)} \quad = f^{(n)}(x) \quad = \frac{d^n y}{dx^n} \quad = \text{nth derivative}$$

EXAMPLE 17

Let us take three derivatives implicitly of $x^2 - y^2 = 9$. (Something nice usually happens in the even derivatives, in this case the second.)

$$x^2 - y^2 - 9 = 0 \qquad 2x - 2y \frac{dy}{dx} = 0 \qquad \frac{dy}{dx} = \frac{x}{y}$$

Quotient rule:

$$\frac{d^2y}{dx^2} = \frac{y(1) - x(dy/dx)}{y^2}$$

$$= \frac{y - x(x/y)}{y^2}$$

Multiply every term top and bottom by y.

$$= \frac{y^2 - x^2}{y^3}$$

From original equation $x^2 - y^2 = 9$; therefore $y^2 - x^2 = -9$.

$$= \frac{-9}{y^3}$$

(Nice simplification— easy to take next derivative).

$$\frac{d^3y}{dx^3} = -9(-3)y^{-4}\left(\frac{dy}{dx}\right)$$

$$= 27y^{-4}\left(\frac{x}{y}\right)$$

$$= \frac{27x}{y^5}$$

EXAMPLE 18—

$$\sin(x^2y^3) - x^4y^5 = 0$$

$$\cos(x^2y^3)[(x^2 3y^2\, dy/dx + y^3 2x)] - x^4 5y^4\, dy/dx - y^5 4x^3 = 0$$

the derivative of the sine of a crazy angle is the cosine of that crazy angle times the derivative of the crazy angle— implicitly and with the product rule

the product rule; the minus sign in front means both parts of the product are negative.

Solving for dy/dx in one step, we get

$$dy/dx = \frac{4x^3y^5 - 2xy^3\cos(x^2y^3)}{3x^2y^2\cos(x^2y^3) - 5x^4y^4}$$

The only difference in the algebra is to multiply out the terms in parentheses in your head (the cosine multiplies each term). The example is then virtually the same as Example 14.

ANTIDERIVATIVES AND DEFINITE INTEGRALS

We are interested in the *antiderivative.* That is, given a function f(x), the antiderivative of f(x), F(x), is a function such that F'(x) = f(x). We are going to explore methods of getting F(x).

The big problem in antiderivatives is that there is no product rule and no quotient rule. You might say, "Hooray! No rules to remember!" In fact, this makes antiderivatives much more difficult to find, and for many functions we are unable to take the antiderivatives. However, in calculus I, antiderivatives are very gentle. Only later do they get longer and more difficult.

RULE I
If F'(x) = G'(x), then F(x) = G(x) + C. If the derivatives are equal, the original functions differ by a constant. In other words, if you have one antiderivative, you have them all; you just have to add a constant.

EXAMPLE 19, SORT OF—

$$F(x) = x^2 + 7 \qquad G(x) = x^2 - 3$$

F'(x) = 2x = G'(x). The difference between F(x) and G(x) is a constant, 10.

RULE 2
If f'(x) = x^N with N ≠ −1, then

$$f(x) = \frac{x^{N+1}}{N+1} + C$$

Add one to the original exponent and divide by the new exponent plus a constant.

RULE 3
If y' = kz', then y = kz + C.

RULE 4

If $y' = f' + g'$, then $y = f + g + C$.

EXAMPLE 20—

$dy/dx = 3x^4 - 8x^3 + 3x + 5$ $(5 = 5x^0)$

$$y = \frac{3x^5}{5} - \frac{8x^4}{4} + \frac{3x^2}{2} + \frac{5x^1}{1} + C$$

$$y = \frac{3x^5}{5} - 2x^4 + \frac{3x^2}{2} + 5x + C$$

EXAMPLE 21—

$dy/dt = \frac{4}{\sqrt{t}} - \frac{8}{t^5} + a = 4t^{-1/2} - 8t^{-5} + a$ $(a = \text{constant})$

$$y = \frac{4t^{1/2}}{\frac{1}{2}} - \frac{8t^{-4}}{-4} + at + C$$

$$= 8t^{1/2} + \frac{2}{t^4} + at + C$$

EXAMPLE 22—

If $y' = 3x^2 + 4x$ and $y = 2$ when $x = 3$, find y.

$$y = \frac{3x^3}{3} + \frac{4x^2}{2} + C = x^3 + 2x^2 + C$$

Now $x = 3$, $y = 2$ to find C.

$2 = (3)^3 + 2(3)^2 + C$. $2 = 27 + 18 + C$. So $C = -43$.

$y = x^3 + 2x^2 - 43$.

RULE 5

If $y' = u^n \, du/dx$ with $n \neq -1$, then $y = \frac{u^{n+1}}{n+1} + C$.

NOTE

Put $u = x$ in rule 5 and you will have rule 2. We will discover that when $N = -1$, the antiderivative is a logarithm.

EXAMPLE 23

Find y if $y' = x(x^2 + 1)^{99}$.

We could multiply this out. On the grounds of sanity, we reject this technique.

Let $u = (x^2 + 1)$. $du/dx = 2x$.

$$y' = x(x^2 + 1)^{99} = \tfrac{1}{2}(x^2 + 1)^{99}(2x) = \tfrac{1}{2}u^{99} \, du/dx$$

$$y = \frac{1}{2}\left(\frac{u^{100}}{100}\right) + C = \frac{(x^2 + 1)^{100}}{200} + C$$

EXAMPLE 24

If $\dfrac{dy}{dx} = \dfrac{x^3}{(x^4 + 11)^6}$, find y.

Let $u = x^4 + 11$. $du/dx = 4x^3$.

$$\frac{dy}{dx} = \frac{1}{4}(x^4 + 11)^{-6}(4x)^3 = \frac{1}{4}u^{-6}\frac{du}{dx}$$

$$y = \frac{1}{4}\left(\frac{u^{-5}}{-5}\right) + C = \frac{-1}{20(x^4 + 11)^5} + C$$

EXAMPLE 25

If $v = 6t^2 + 4t + 3$ and $s = 40$ when $t = 1$, find s. s = distance, v = velocity, t = time, v = ds/dt.

$ds/dt = 6t^2 + 4t + 3$

$$s = \frac{6t^3}{3} + \frac{4t^2}{2} + 3t + C$$

$$s = 2t^3 + 2t^2 + 3t + C$$

Use $t = 1$ when $s = 40$ to get $40 = 2(1)^3 + 2(1)^2 + 3(1) + C$. $C = 33$.

$s = 2t^3 + 2t^2 + 3t + 33$

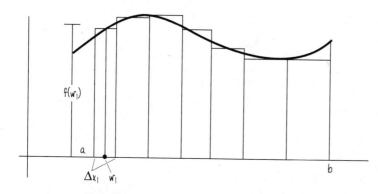

The development of the area as motivation for the definite integral is detailed in most calculus books. We will sketch the development.

A. Given the region $y = f(x)$, $x = a$, $x = b$, x axis.

B. Divide the interval [a,b] into n intervals, Δx_1, Δx_2, Δx_3, . . . Δx_n. Δx_i represents an arbitrary interval.

C. Let w_i be any point in the interval Δx_i.

D. Δx_1 represents the width of the first approximating rectangle. $f(w_1)$ represents the height of the first rectangle. $\Delta x_1\, f(w_1)$ represents the area of the first approximating rectangle.

E. Do this for all the rectangles. We get

$$f(w_1)\Delta x_1 + f(w_2)\Delta x_2 + f(w_3)\Delta x_3 + \cdots + f(w_n)\Delta x_n = \sum_{i=1}^{n} f(w_i)\Delta x_i$$

$$\lim_{\substack{n \to \infty \\ \text{all } \Delta x_i \to 0}} \sum_{i=1}^{n} f(w_i)\Delta x_i = \int_{a}^{b} f(x)\ dx$$

The definite integral.

In the above example, the definite integral represents the area. As in the case of derivatives, we would like the rules so that it would be unnecessary to do this process of approximating rectangles.

NOTE
If the velocity = v(t),

$$\int_a^b v(t)\, dt$$

represents the distance traveled from time a to time b.

RULE 6

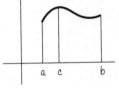

$$\int_a^b 1\, dx = b - a$$

RULE 7

$$\int_a^b cf(x) = c \int_a^b f(x)\, dx$$

RULE 8

$$\int_a^b (f(x) + g(x))\, dx = \int_a^b f(x)\, dx + \int_a^b g(x)\, dx$$

RULE 9

$$\int_a^b f(x)\, dx = \int_a^c f(x)\, dx + \int_c^b f(x)\, dx$$

RULE 10

$$\int_a^a f(x)\, dx = 0$$

RULE 11

$$\int_a^b f(x)\, dx = -\int_b^a f(x)\, dx$$

RULE 12
The fundamental theorem of integral calculus. If f(x) is continuous on [a,b] and F(x) is any antiderivative of f(x), then

$$\int_a^b f(x)\, dx = F(b) - F(a).$$

EXAMPLE 26—

$$\int_1^4 (3x^2 + 6)\, dx$$

$$= \int_1^4 3x^2\, dx + 6 \int_1^4 1\, dx$$

$$= x^3 + 6x \Big[_1^4 = [4^3 + 6(4)] - [1^3 + 6(1)] = 81$$

NOTE

$\int_a^b f(x)\, dx$, the definite integral, is a number if a and b are numbers. $\int f(x)\, dx$ the antiderivative or indefinite integral, is a family of functions, each of which differs from the others by a constant.

We rewrite previous rules using the indefinite integral.

$$\int x^n\, dx = \frac{x^{n+1}}{n+1} + C \qquad n \neq -1$$

$$\int [cf(x) + g(x)]\, dx = c \int f(x)\, dx + \int g(x)\, dx$$

$$\int u^N \frac{du}{dx} \cdot dx = \frac{u^{N+1}}{N+1} + C \qquad N \neq -1$$

EXAMPLE 27—

$$\int \frac{(x^4 + 3x)^2}{x^5}\, dx$$

$$\frac{(x^4 + 3x)^2}{x^5} = \frac{x^8 + 6x^5 + 9x^2}{x^5} = \frac{x^8}{x^5} + \frac{6x^5}{x^5} + \frac{9x^2}{x^5}$$

$$= x^3 + 6 + 9x^{-3}$$

$$\int \frac{(x^4 + 3x)^2\, dx}{x^5} = \int (x^3 + 6 + 9x^{-3})\, dx = \frac{x^4}{4} + 6x + \frac{9x^{-2}}{-2} + C$$

$$= \frac{x^4}{4} + 6x - \frac{9}{2x^2} + C$$

EXAMPLE 28—

This is a change of variables. We let u equal whatever is in parentheses, under a radical sign, etc. Let $u = x^3 + 1$. $du/dx = 3x^2$. Solve for dx. $du = 3x^2 \, dx$. So $dx = du/3x^2$.

$$\int_0^3 x^2(x^3 + 1)^{1/2} \, dx$$

The limits must change. $u = x^3 + 1$. $x = 0$. $u = 0^3 + 1 = 1$.

$$x = 3. \ u = 3^3 + 1 = 28.$$

$$\int_{x=0}^{x=3} x^2(x^3 + 1)^{1/2} \, dx = \int x^2 u^{1/2} \, du/3x^2$$

$$= \int_{u=1}^{u=28} (1/3)u^{1/2} \, du$$

$$= (1/3)u^{3/2}/(3/2) \Big[\Big._{1}^{28}$$

$$= (2/9)u^{3/2} \Big[\Big._{1}^{28} = (2/9)(28^{3/2} - 1)$$

EXAMPLE 29—

Here's a tricky one.

$$\int_1^2 \sqrt{x^4 - x^2} \, dx \qquad \sqrt{x^4 - x^2} = \sqrt{x^2(x^2 - 1)} = x\sqrt{x^2 - 1}$$

Let $u = x^2 - 1$. $x = 1$; $u = 0$. $x = 2$; $u = 3$. $dx = du/2x$.

$$\int_{x=1}^{x=2} \sqrt{x^4 - x^2} \, dx = \int_{x=1}^{x=2} x(x^2 - 1)^{1/2} \, dx = \int xu^{1/2} \, du/2x$$

$$= \int_{u=0}^{u=3} \tfrac{1}{2}u^{1/2} \, du = \tfrac{1}{2}u^{3/2}/(3/2) \Big[\Big._{u=0}^{u=3}$$

$$= (1/3)u^{3/2} \Big[\Big._{u=0}^{u=3} = (1/3)3^{3/2} - 0$$

$$= (1/3)(3)(3)^{1/2} = 3^{1/2}$$

NOTE

If you make a u substitution and the x's do not cancel out, either a new technique must be used or the problem cannot be solved.

EXAMPLE 30—

$\int \sin 5x \, dx$

Let $u = 5x$. $dx = du/5$.

$\int \sin 5x \, dx = \int \sin u \, du/5$

$= (1/5) \int \sin u \, du = (1/5)(-\cos u) + C$

$= -\cos (5x)/5 + C$

The integral of the sine is minus cosine and the integral of the cosine is the sine.

You must replace u with 5x since the problem originally had x.

NOTE

You must know how to do an integral like this one by sight, because if you don't, some of the calc II integrals will become virtually endless.

EXAMPLE 31—

$\int_{x=0}^{\pi/12} \tan^4 3x \sec^2 3x \, dx$

You must remember that the derivative of the tangent is the secant squared.

Let $u = \tan 3x$. $du = 3 \sec^2 3x \, dx$. $du/3 = \sec^2 3x \, dx$.

$x = 0$; $\tan (0) = 0$. So $u = 0$.

$x = \pi/12$. $\tan (3\pi/12) = \tan (\pi/4) = 1$. So $u = 1$.

$\int_{x=0}^{\pi/12} \tan^4 3x \sec^2 3x \, dx = \int_{u=0}^{u=1} (1/3)u^4 \, du$

$= (1/3)u^5/5 \Big|_0^1 = (1/15)u^5 \Big|_0^1 = 1/15$

Finding the Area Under the Curve Using the Definition of the Definite Integral

One of the most laborious tasks is to find the area using the definition. Doing one of these problems will make you forever grateful that there are some rules for antiderivatives, especially the fundamental theorem of calculus.

EXAMPLE 32—

Find $\int_3^6 (x^2 + 4x + 7)\, dx$ using the definition of the definite integral.

Before we start, we need two facts:

$$1 + 2 + 3 + \cdots + n = n(n + 1)/2$$

$$1^2 + 2^2 + 3^2 + \cdots + n^2 = n(n + 1)(2n + 1)/6$$

These formulas for the sum of the first n positive integers and the sum of the squares of the first n positive integers can be found in some precalculus books but are not too easily proved.

Now we are ready—not happy—but ready to start.

1. Divide the interval $3 \le x \le 6$ into n equal parts. The left end $3 = x_0$, $x_1 = 3 + 1\, \Delta x$, $x_2 = 3 + 2\, \Delta x$, $x_3 = 3 + 3\, \Delta x, \ldots$, and $x_n = 3 + n\, \Delta x = 6$. Solving for Δx, $\Delta x = (6 - 3)/n = 3/n$.

2. From before, the approximate sum is $f(w_1)\Delta x_1 + f(w_2)\Delta x_2 + f(w_3)\Delta x_3 + \cdots + f(w_n)\Delta x_n$. All of the Δx_i are equal to Δx, and we will take the right end of each interval as the point where we will take the height. Therefore $w_1 = x_1$, $w_2 = x_2, \ldots$, and $w_n = x_n$.

3. Rewriting 2, we factor out the Δx and get $[f(x_1) + f(x_2) + f(x_3) + \cdots + f(x_n)]\Delta x$.

4. Now $f(x) = x^2 + 4x + 7$.

   $$f(x_1) = (3 + 1\, \Delta x)^2 + 4(3 + 1\, \Delta x) + 7$$

   $$f(x_2) = (3 + 2\, \Delta x)^2 + 4(3 + 2\, \Delta x) + 7$$

$$f(x_3) = (3 + 3\ \Delta x)^2 + 4(3 + 3\ \Delta x) + 7$$

$$f(x_n) = (3 + n\ \Delta x)^2 + 4(3 + n\ \Delta x) + 7$$

5. Our task is now to add all these up and then multiply everything by Δx.

   A. If we multiply out $f(x_1)$, we see that the number we get from this term is $3^2 + 4(3) + 7 = 28$. We see that every term, if we were to multiply them out, would have a 28. Since there are n terms, the sum would be 28n.

   B. Let's look at the Δx terms. $f(x_1)$ gives us $6(1\ \Delta x) + 4(1\ \Delta x) = 10(1\ \Delta x)$. $f(x_2)$ gives us $6(2\ \Delta x) + 4(2\ \Delta x) = 10(2\ \Delta x)$. Similarly, $f(x_3) = 10(3\ \Delta x)$. And $f(x_n) = 10(n\ \Delta x)$. Adding and factoring, we get $10\ \Delta x(1 + 2 + 3 + \cdots + n)$.

   C. Looking at the $(\Delta x)^2$ terms, we would get

   $$1^2(\Delta x)^2 + 2^2(\Delta x)^2 + 3^2(\Delta x)^2 + \cdots + n^2(\Delta x)^2$$

   Factoring, we would get

   $$(\Delta x)^2(1^2 + 2^2 + 3^2 + \cdots + n^2)$$

6. Now, adding everything up, multiplying by Δx, hoping everything fits on one line, we get

   $$[28n + 10\ \Delta x\ (1 + 2 + 3 + \cdots + n) + (\Delta x)^2(1^2 + 2^2 + 3^2 + \cdots + n^2)]\Delta x$$

7. Substituting the formulas in the beginning and remembering $\Delta x = 3/n$, we get

   $$\{28n + 10(3/n)[n(n + 1)/2] + (3/n)^2\ [n(n + 1)(2n + 1)/6]\}(3/n)$$

8. Using the distributive law, we get three terms:

   A. $(28n)(3/n) = 84$.

   B. $10(3/n)[n(n + 1)/2](3/n) = 45(n + 1)/n$
      $\lim\limits_{n \to \infty}$ of B $= 45(1) = 45$

C. $(9/n^2)[n(n + 1)(2n + 1)/6](3/n) = [(27/6)(2n^2 + 3n + 1)/6](n^2)$

$\lim\limits_{n \to \infty} C = (27/6)(2) = 9$

9. Adding $A + B + C$, we get a value for the area of $84 + 45 + 9 = 138$.

NOTE

By letting n go to infinity, we are doing two things: chopping up the interval $3 \le x \le 6$ into more and more rectangles and, since $\Delta x = 3/n$, making each rectangle narrower and narrower.

Wow!!!!!!!!! Are we grateful for the fundamental theorem!

$$\int_3^6 (x^2 + 4x + 7)\, dx = \frac{x^3}{3} + 2x^2 + 7x \Big|_3^6$$

$$= \left[\frac{6^3}{3} + 2(6)^2 + 7(6)\right] - \left[\frac{3^3}{3} + 2(3)^2 + 7\cdot(3)\right]$$

$$= (72 + 72 + 42) - (9 + 18 + 21)$$

$$= 186 - 48 = 138!$$

Finally we will learn, sadly, that most integrals cannot be done. As we go on in math, we will learn many approximation methods. Also, we will learn how accurate these approximations are. This is OK because we don't live in a perfect world—or is this a surprise to you? As we go on, the methods of approximation will become more involved. Let us take a look at some crude ones. We will approximate

$$\int_{-1}^5 (x^2 + 2)\, dx$$

in three ways.

EXAMPLE 33

Approximate this integral with three equal subdivisions, using the right end of each one. Here's the picture:

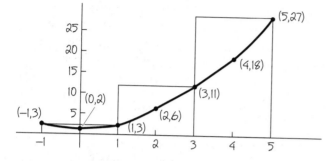

The approximate area is $f(w_1)\Delta x_1 + f(w_2)\Delta x_2 + f(w_3)\Delta x_3$. Each $\Delta x = 2$, and $w_1 = 1$, $w_2 = 3$, and $w_3 = 5$, the right ends of each interval. The approximation is $(\Delta x)[f(1) + f(3) + f(5)] = 2(3 + 11 + 27) = 82$.

EXAMPLE 34—

Same picture, same intervals, the minimum approximation, the smallest value in each interval, S_3.

$S_3 = \Delta x[f(0) + f(1) + f(3)] = 2(2 + 3 + 11) = 32$

EXAMPLE 35—

Same picture, $x_0 = -1$, $x_1 = 0$, $x_2 = \frac{1}{2}$, $x_3 = 1$, $x_4 = 5$; midpoints.

The approximation is

$f(w_1)(\Delta x_1) + f(w_2)(\Delta x_2) + f(w_3)(\Delta x_3) + f(w_4)(\Delta x_4)$

$= f(-\tfrac{1}{2})(1) + f(\tfrac{1}{4})(\tfrac{1}{2}) + f(\tfrac{3}{4})(\tfrac{1}{2}) + f(3)(4)$

$= \left(\dfrac{9}{4}\right)(1) + \left(\dfrac{21}{16}\right)(\tfrac{1}{2}) + \left(\dfrac{29}{16}\right)(\tfrac{1}{2}) + 11(4)$

$= 48\tfrac{9}{16}$

Of course we do not have to approximate this integral, since it is easily done. However, we would need to approximate, say, this one:

$$\int_{-1}^{5} \frac{1}{1 + x^4}\, dx$$

A theorem that is now mentioned much more often in calc I is the average value theorem. It says that if we have an integrable function on the interval $a \leq x \leq b$, there exists a point c between a and b such that

$$f(c) = \frac{1}{b-a} \int_a^b f(x) \, dx$$

We will demonstrate by picture.

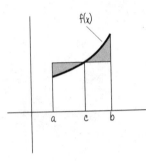

Suppose we have the function as pictured. There is a point c where the two shaded areas are the same. Fill in the top one in the bottom space. Thus the area of the rectangle equals the area under the curve. Area of the rectangle is base times height. Base = b – a. Height = f(c).

$$f(c)\,(b-a) = \int_a^b f(x) \, dx$$

Now divide by b – a.

That's it. Let's do an example.

EXAMPLE 36—

Find the average value for $f(x) = x^2$, $2 \leq x \leq 5$.

$$\frac{1}{5-2} \int_2^5 x^2 \, dx = x^3/3 \Big|_2^5 = \frac{1}{9}(5^3 - 2^3) = \frac{1}{9}(117) = 13$$

If I actually wanted to find the point c in the theorem that gives the average value, $f(c) = c^2 = 13^{1/2} = 3.6$, which clearly is between 2 and 5.

Let's do a word problem.

EXAMPLE 37—

During the 12 hours of daylight, the temperature Fahrenheit is given by $T = 60 + 4t - t^2/3$. Find the average temperature over the 12-hour period. The average temperature value

$$= \frac{1}{12 - 0} \int_0^{12} \left(60 + 4t - \frac{t^2}{3} \right) dt$$

$$= \frac{1}{12} \left(60t + 2t^2 - \frac{t^3}{9} \right)$$

$$= \frac{60(12)}{12} + \frac{2(12)(12)}{12} - \frac{(12)(12)(12)}{(3)(3)(12)}$$

$= 68°F \ (20°C)$, a delightful average temperature.

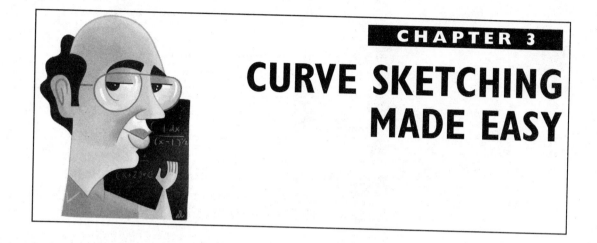

CHAPTER 3

CURVE SKETCHING MADE EASY

The topic I think I can teach better than anyone else in the world is this one. The only question is whether I could write it down. I think I did, and I think you'll really enjoy it!

Since we can spend an almost infinite amount of time on the topic, we will restrict our discussion to polynomials and rational functions (polynomials over polynomials), except for a few examples at the end.

TERMS AND SPECIAL NOTATIONS

1. For curve-sketching purposes, we define an asymptote as a line to which the curve gets very close at the end but never hits. All your life you have been told a curve cannot hit an asymptote. This is wrong. An asymptote is a straight-line approximation to a curve near its end, that is, when x or y goes to plus or minus infinity. In the middle of the curve, the curve is not a straight line and can hit the asymptote. The x axis is an asymptote although the curve hits the axis five

times. At the end of the curve, the curve gets close to the axis but does not hit it.

2. |f(3)| = infinity. As x gets close to 3, f(x) gets very, very big (heading to plus infinity) or very, very small (heading to minus infinity).

3. f(6⁺). Substitute a number a little larger than 6, such as 6.01.

4. f(6⁻). Substitute a number a little smaller than 6, such as 5.99.

Our first goal is to sketch, in under two minutes, curves like

$$y = f(x) = \frac{(x-2)^6(x+1)^7}{(x+2)^3(x-1)^{20}}$$

Yes, not only is it possible, but almost all of my students do it and so will you!!!!!

INTERCEPTS

x intercepts. Just like a straight line, x intercept means a point where y = 0. If we have a fraction, y = 0 means the top of the fraction = 0.

EXAMPLE 1—

$$y = \frac{x^4(2x-3)^7(x+4)^8}{(x+2)^3(x-1)^{20}}$$

y = 0 means the top is 0. "Top is 0" means x = 0, 2x − 3 = 0, or x + 4 = 0. You ignore the exponents, since $x^4 = 0$ means x = 0. x = 0, 3/2, and −4. The intercepts are (0,0), (1.5,0), and (−4,0).

EXAMPLE 2—

$$y = \frac{x^2 - 2x - 3}{x}$$

Factor the top.

$(x - 3)(x + 1) = 0$. Intercepts are $(3,0)$ and $(-1,0)$.

EXAMPLE 3—

$y = x^4(x - 1)^5(x - 2)^6$.

Intercepts are $(0,0)$, $(1,0)$, and $(2,0)$.

   y intercepts. Just like the line, y intercept means a point where $x = 0$.

EXAMPLE 4—

$$y = \frac{(x + 1)^8(x - 2)^5}{(x - 1)^7(x + 4)}$$

Substitute x = 0.

$\dfrac{(1)^8(-2)^5}{(-1)^7(4)} = 8$. y intercept is $(0,8)$.

EXAMPLE 5—

$$y = \frac{x^2 - 2x - 3}{x^2}$$

For $x = 0$, we get $-3/0$. There is no y intercept.
   Warnings:

1. If you get the *sign* of the y intercept wrong, you will *never, never* sketch the curve properly.

2. Functions have one y intercept at most (one or none).

3. If we have the intercept $(0,0)$, it is one of the x intercepts, maybe the only one, but the only y intercept. We do not have to waste time trying to find another one!

VERTICAL ASYMPTOTES

A rational function has a vertical asymptote whenever the bottom of the fraction is equal to 0.

EXAMPLE 6—

$$y = \frac{x + 31}{x^4(x - 4)^6(x + 3)^5}$$

Asymptotes are vertical lines $x = 0$, $x = 4$, and $x = -3$.

Two more definitions need to be reviewed before we move on.

DEFINITION A

Degree—if a polynomial has one variable, it is the highest exponent.

DEFINITION B

Leading coefficient—the coefficient of the highest power.

EXAMPLE 7—

$$y = 4x^3 - 7x^6 + 2$$

Degree is 6. Leading coefficient is −7.

>
> **NOTE**
>
> Since polynomials have no denominators, they have no vertical asymptotes. As we will see, they have no asymptotes at all.

HORIZONTAL ASYMPTOTE TYPE I

(Don't be scared. There are only two types!)

Suppose $y = P(x)/Q(x)$. P and Q are polynomials. If the degree of P (top) is less than the degree of Q (bottom), the horizontal asymptote is $y = 0$, the x axis.

EXAMPLE 8—

$$y = \frac{3x^2 - 7x}{8 + 5x^4} = \frac{\dfrac{3x^2}{x^4} - \dfrac{7x}{x^4}}{\dfrac{8}{x^4} + \dfrac{5x^4}{x^4}} = \frac{3/x^2 - 7/x^3}{8/x^4 + 5}$$

As x goes to infinity, $3/x^2$, $-7/x^3$, and $8/x^4$ all go to 0. So $y = (0 - 0)/(0 + 5) = 0$. The asymptote is $y = 0$!!!

NOTE 1

We divided by x^4, that is, x to the highest power.

NOTE 2

When you do the problem, don't actually do this. Since you know anytime the degree of the top is smaller than the degree of the bottom you get $y = 0$, just use $y = 0$ when this happens. Easy, isn't it?!!

HORIZONTAL ASYMPTOTE TYPE 2

EXAMPLE 9—

$$y = \frac{6x^3 + 2}{5x - 7x^3}$$

Both degrees are 3. If the degree of the top equals the degree of the bottom, the horizontal asymptote is $y = a/b$, where a is the leading coefficient of the top and b is the leading coefficient of the bottom. Asymptote is $y = 6/(-7)$. Let us verify.

$$y = \frac{6x^3 + 2}{5x - 7x^3} = \frac{(6x^3)/x^3 + 2/(x^3)}{(5x)/(x^3) - (7x^3)/(x^3)} = \frac{6 + 2/(x^3)}{(5)/(x^2) - 7}$$

As x goes to infinity, $2/x^3$ and $5/x^2$ go to 0. Asymptote is $y = -6/7$.

OBLIQUE (SLANTED LINE) ASYMPTOTE

This occurs when the degree of the top is exactly 1 more than the bottom.

EXAMPLE 10—

$$y = \frac{x^2 - 2x - 3}{x + 4}$$

Degree of the top = 2; degree of the bottom = 1. Oblique asymptote.

We must, unfortunately, long divide the bottom into the top. If you know it, use synthetic division.

$$x + 4 \overline{\smash)\begin{array}{r} x - 6 + 21/(x+4) \\ x^2 - 2x - 3 \end{array}} \qquad y = \frac{x^2 - 2x - 3}{x + 4} = x - 6 + \frac{21}{x + 4}$$
$$\underline{x^2 + 4x} $$
$$-6x - 3$$
$$\underline{-6x - 24}$$
$$21$$

As x goes to infinity, the remainder $21/(x + 4)$ goes to 0. The oblique asymptote is $y = x - 6$.

NOTE I

If the degree of the top is more than the bottom but not 1, there are no oblique asymptotes.

NOTE 2

At most there is one oblique asymptote or one horizontal asymptote, but not both. There might be neither.

CURVE SKETCHING BY THE PIECES

Before we take a long example, we will examine each piece. When you understand each piece, the whole will be easy.

EXAMPLE II—

$$f(x) = 3(x - 4)^6$$

The intercept is (4,0). We would like to know what the curve looks like near (4,0). Except at the point (4,0), we do not care what the exact value is for y, which is necessary in an exact graph. In a sketch we are only interested in the *sign* of the y values. We know f(4) = 0. f(3.9) = 3(3.9 − 4)^6 = 0.000003. We don't care about its value. We only care that it is positive. Our notation will be f(4$^-$) is positive. Similarly f(4$^+$) is positive. What must the picture look like? At x = 4, the value is 0. To the left and to the right, f(x) is positive.

The picture looks like this where a = f(4$^-$) and b = f(4$^+$) . . .

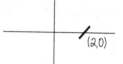

EXAMPLE 12—

f(x) = −7(x − 6)^{100}

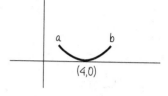

Intercept (6,0). f(6$^-$) is negative. f(6$^+$) is negative. The picture is . . .

To summarize, if the exponent is *even positive,* the sketch does *not* cross at the intercept.

EXAMPLE 13—

f(x) = 6(x − 2)^{11}

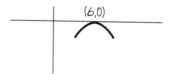

Intercept (2,0), f(2$^-$) is *negative;* f(2$^+$) is *positive.* The curve around (2,0) looks like . . .

EXAMPLE 14—

f(x) = −7(x + 4)^{3743}

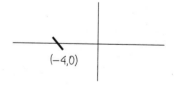

Intercept (−4,0), f(−4$^-$) is *positive;* f(−4$^+$) is *negative.* Around (−4,0) the curve will look like . . .

To summarize, if the exponent is *odd positive,* the sketch *will* cross at the intercept.

Let's see what it will look like near the vertical asymptotes. We attack the problem in exactly the same way.

EXAMPLE 15—

$$f(x) = \frac{7}{(x − 4)^8}$$

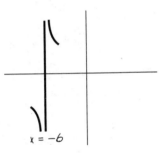

Vertical asymptotes at x = 4, f(4⁻), f(4⁺) are positive, and the curve near x = 4 looks like . . .

To reemphasize, $f(4^+) = f(4.1) = 7/(4.1 - 4)^8 = 7/0.00000001 = 700,000,000$, which is *big*. The curve tends to plus infinity, from the right side of 4. Similarly, the curve goes to plus infinity from the left side.

EXAMPLE 16—

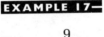

$$f(x) = \frac{-7}{(x - 3)^{100}}$$

Vertical asymptote at x = 3. f(3⁻) and f(3⁺) are negative. The curve near x = 3 looks like . . .

To summarize, if the exponent is *even positive* in the denominator, near the vertical asymptote, *both ends go to either plus infinity or minus infinity.*

EXAMPLE 17—

$$f(x) = \frac{9}{(x + 6)^{11}}$$

Asymptote x = –6. f(–6⁻) is negative; f(–6⁺) is positive. The curve is . . .

EXAMPLE 18—

$f(x) = -3/(x - 5)$

Exponent is 1 (odd). Asymptote x = 5. f(5⁻) is positive; f(5⁺) is negative. The curve near x = 5 is . . .

To summarize, if the exponent is *odd positive* in the denominator, on one side of the asymptote the curve goes to plus infinity and on the other the curve goes to minus infinity.

We are now ready to put all the pieces together. With some study and a little practice, you positively will be

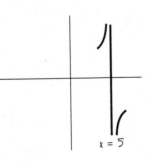

able to sketch curves with intercepts and asymptotes only in under two minutes!!!!!!!

EXAMPLE 19—

$$f(x) = \frac{(x-2)^6(x+1)^7}{(x+2)^3(x-1)^{20}}$$

First locate the intercepts. x intercept means y = 0 means top of fraction = 0. (2,0) and (–1,0). y intercept, the only long part. Substitute x = 0.

$$f(0) = \frac{(-2)^6(1)^7}{(2)^3(-1)^{20}} = 8$$

The intercept is (0,8). Vertical asymptotes: bottom of fraction = 0. x = –2 and x = 1. Horizontal asymptote: if we multiplied out the top (something you would *never* do), the highest power of x on top is x^{13}. On the bottom is x^{23}. The degree of the top is less than the bottom. The horizontal asymptote is y = 0. Oblique asymptote: none, since there is a horizontal one.

We are ready to start the sketch. It is advisable to use three colors, one for the axes, one for the asymptotes, and one for the sketch.

We need to substitute only one number!!! That number is to the right of the rightmost vertical asymptote or x intercept. $f(2^+)$ is positive. Since the power of (x – 2) is even, namely 6, the curve does not cross at (2,0). So far the sketch is . . .

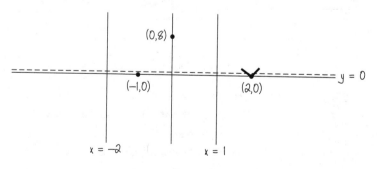

The sketch now heads for the asymptote $x = 1$. It must go to plus infinity, since if it went to minus infinity, there would have to be another x intercept between 1 and 2 and there isn't. The power of $(x - 1)^{20}$ is even. For an asymptote, that means both ends are in the same location. Since one part is at plus infinity, so is the other. The curve now heads through (0,8) toward (−1,0). It looks like . . .

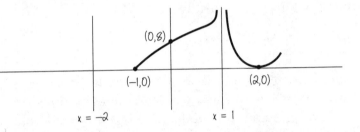

Since the power of $(x + 1)$ is odd, 7, there is a crossing at (−1,0). The sketch heads to minus infinity at $x = -2$. Since the power of $(x + 2)$ is odd and an odd power means one end at plus infinity and one at minus infinity, one end is already at minus infinity, so that on the other side the curve goes to plus infinity. *Remember,* both ends head for the horizontal asymptote $y = 0$, the x axis. The sketch can now be finished. . . .

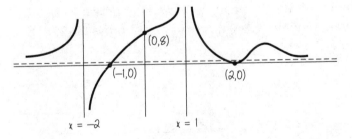

Well, that's it. With a little practice you'll be like a pro! We'll do more examples, of course. You should practice the ones in this book.

Before we do other examples, let us look at the right end of the curve that we drew above.

In the top figure, can we find out how high M is? Yes! Later in this section we will.

In the middle, could this be the right end? Perhaps, but we don't have enough info to know what the end looks like yet. In some very complicated cases, we might not ever be able to determine what the end is like (exactly).

In the bottom, could this be the right end? No!! There would have to be another intercept after (2,0), and we know there aren't any.

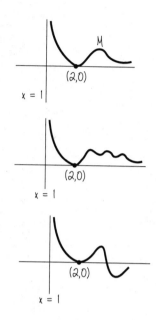

EXAMPLE 20—

$$f(x) = \frac{3x^4(x-4)^5(x-8)^6}{2(x+3)^7(x+6)^6(x-2)^2}$$

x intercepts: (8,0), (4,0), and (0,0), which is also the y intercept. Vertical asymptotes: $x = 2$, $x = -3$, $x = -6$. Horizontal asymptote: By inspection, the leading term on the top is $3x^{15}$ and the leading term on the bottom is $2x^{15}$. Same degree. The asymptote is $y = 3/2$, the leading coefficient on top over leading coefficient on bottom.

Starting the sketch at (8,0), $f(8^+)$ is positive. The power of $(x-8)$ is even, so there is no crossing. The sketch starts . . .

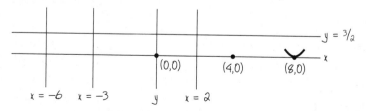

From the sketch, $f(4^+)$ is positive. Since $(x-4)$ has odd power, there is a crossing at (4,0) heading down to minus infinity at x = 2. The power of (x – 2)² is even. So one end at minus infinity means both ends are there, heading up to (0,0). x⁴ is an even power, so that there is no crossing, and the sketch continues . . .

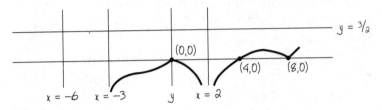

We head to minus infinity at x = –3. (x + 3)⁷ is an odd power. So since one end is at minus infinity, the other is at plus infinity. It goes down (we don't know how far) but never hits the x axis and heads back to plus infinity at x = –6. Since (x + 6)⁶ is an even power, both ends are at plus infinity. The *ends* both go to y = ¾, the horizontal asymptote. The sketch can now be finished. . . .

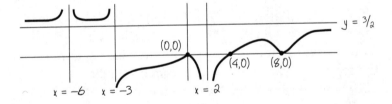

You should be getting better now. Let's try an oblique asymptote.

EXAMPLE 21—

f(x) = (x² – 4x + 4)/(x – 1)

Since the degree of the top is 1 more than the bottom, we have an oblique asymptote. We need three forms of the equation: the original, the factored, and the divided.

f(x) = (x² – 4x + 4)/(x – 1) = (x – 2)²/(x – 1)

= x – 3 + 1/(x – 1)

x intercept—top = 0 in the second form: (2,0). y intercept—x = 0, easiest found in the first form: (0,–4). Vertical asymptote—first or second form: x = 1. Oblique
asymptote—third form: y = x – 3 with remainder going
to 0.

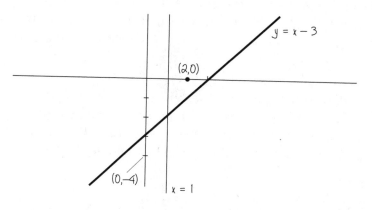

Again we look at the rightmost vertical asymptote or
x intercept, in this case (2,0). f(2⁺) (form 2) is positive.
(x – 2)² is an even power, so there is no crossing, heading up to plus infinity at x = 1. Since the power of x –
1 (1) is odd, the other end is at minus infinity. The
sketch then goes through the point (0,–4) with both
ends going to the line y = x – 3. The sketch is . . .

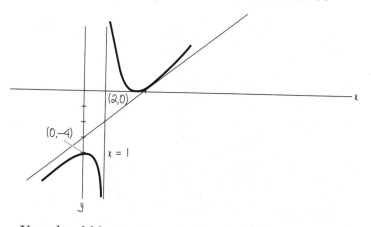

You should be getting *a lot better* now!! Let us try a
polynomial, to show you how easy it is.

EXAMPLE 22

$f(x) = x^4(x - 1)^5(x - 2)^6(x - 3)^7(5 - x)^9$

Intercepts: (0,0), (1,0), (2,0), (3,0), and (5,0). No asymptotes. Watch how easy this is! $f(5^+)$ is negative. $(5 - x)^9$ is an odd power, so cross. $(x - 3)^7$ is an odd power, so cross. $(x - 2)^6$ is an even power, no cross. $(x - 1)^5$ is an odd power, so cross. x^4 is an even power, no cross. The leading term, $-x^{31}$, dominates when x is big (say x = 100) or when x is small (say x = −100). $-(100)^{31}$ negative—right end goes to minus infinity. $-(-100)^{31}$ positive—left end goes to plus infinity. Briefly, right to left, looking at the exponents only—cross, cross, no cross, cross, no cross, and the sketch looks like . . .

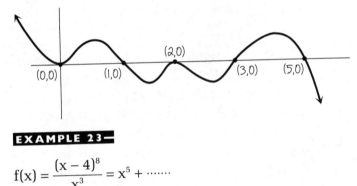

EXAMPLE 23

$f(x) = \dfrac{(x - 4)^8}{x^3} = x^5 + \cdots\cdots$

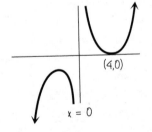

Intercept (4,0). Vertical asymptote x = 0 (y axis). No oblique or horizontal asymptote since degree of top is 5 more than the bottom. $(100)^5$ is positive—right end goes to plus infinity. $(-100)^5$ is negative—left end goes to minus infinity. The sketch is . . . ta-da! . . .

The next area of curve sketching involves maximum points, minimum points, inflection points, and cusps. Since much of this involves derivatives and factoring, more care and time is needed.

DEFINITIONS
Relative maximum (*minimum*)—the largest (smallest) y value in a region. *absolute maximum* (*minimum*)—the largest (smallest) y value of all.

If the sketch goes to plus infinity (minus infinity), all maximums (minimums) are relative. A, C, E, H, and J are relative maximums. B, D, F, G, I, and K, are relative minimums. No absolute maximums. F is the absolute minimum.

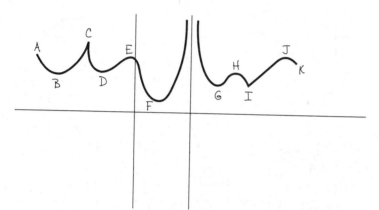

As you can see, there are three kinds of maximums and minimums.

1. If there is a finite domain, the left and right end points (A and K) are sometimes relative maximums or minimums. (They are in most of the examples we do.)

2. *Cusps,* C and I, will be discussed later.

3. *Round maximums* and *minimums* will be discussed now. As you can see, the slope of the tangent line exists at these points and is equal to 0.

Let us proceed.

TESTING FOR ROUND MAXIMUMS AND MINIMUMS

If $y = f(x)$ and there is a round max or min at $x = c$, $f'(c) = 0$.

TEST 1 for round max and min points:

If $f'(c^-)$ is negative and $f'(c^+)$ is positive, we have a minimum.

If $f'(c^-)$ is positive and $f'(c^+)$ is negative, we have a maximum.

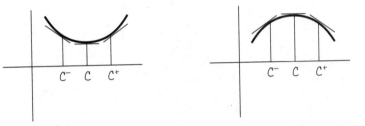

If $f'(c^-)$ and $f'(c^+)$ are the same sign, it is neither a max nor a min.

Believe it or not, at this time we have to define *down* and *up*. A curve is down (up) if the tangent line drawn to the curve is above (below) the curve itself.

Down

Up

DEFINITION

Inflection point—a point at which a curve goes from up to down or down to up.

Down to up   Up to down

A test for inflection points is $f''(c) = 0$. There are in fact two tests to tell whether a point is an inflection point.

TEST 1 (easier)

  A. If $f'''(c) \neq 0$, then c is an inflection point.

  B. If $f'''(c) = 0$, this test fails and you must use another test.

TEST 2

  A. If $f''(c^+)$ and $f''(c^-)$ have different signs, c is an inflection point.

  B. If $f''(c^+)$ and $f''(c^-)$ have the same sign, c is not an inflection point.

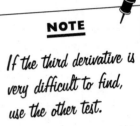

NOTE

If the third derivative is very difficult to find, use the other test.

  There is also a second and easier test for max and min points. Suppose $f'(c) = 0$. If $f''(c)$ is positive, it means the slope is increasing. This means the curve is facing up, which means a *minimum*. Suppose $f'(c) = 0$ and $f''(c)$ is negative. This means the slope is decreasing, the curve faces down, and we have a *maximum*. If $f''(c) = 0$, then we use the other test.

PROBLEM

Before we sketch some more curves, let's make sure we all understand each other. There is a kind of problem my fellow lecturer Dan Mosenkis at CCNY likes to give his students. It's not my cup of tea or cup of anything else, but I think it will help you a lot. We have made up a craaaazy function, $f(x)$. Its picture is on the next page. For each listed value of x, A through K, look at $f(x)$ and estimate the sign of $f(x)$, $f'(x)$, and $f''(x)$ at each point. Enter one of the following symbols on the chart: + (if positive), – (if negative), 0, and ? (if it does not exist).

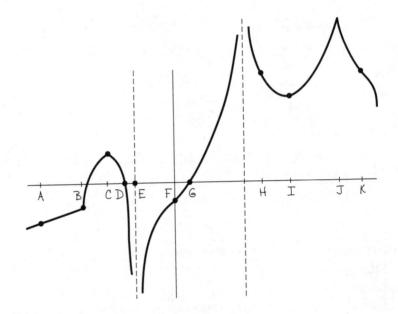

Answer at the end of this chapter!!

x	A	B	C	D	E	F	G	H	I	J	K
f(x)											
f'(x)											
f''(x)											

You are now ready for your first curve sketch involving max and min inflection points.

EXAMPLE 24—

$y = f(x) = x^4 - 4x^3$

Take three derivatives and factor each equation.

$y = x^4 - 4x^3 = x^3(x - 4)$

$y' = 4x^3 - 12x^2 = 4x^2(x - 3)$

$y'' = 12x^2 - 24x = 12x(x - 2)$

$y''' = 24x - 24 = 24(x - 1)$

Intercepts: y = 0 (0,0), (4,0). max and min: y' = 0. x = 0, x = 3. To find the y values, you must substitute each x value *into the original equation,* since that is the curve we are sketching. We get the points (0,0) and (3,−27).

NOTE
It is almost always easier to find the y value from the factored equation of y.

We will use both tests to test points. However, when you do the problem, once you get an answer for one point, go on to the next point.

We first substitute x = 0,3 into f''(x) = 12x(x − 2). f''(0) = 0, so this test fails. f''(3) is positive, so (3,−27) is a minimum.

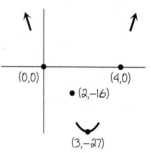

TEST 2 f'(x) = 4x²(x − 3). f'(0⁺) is negative and f'(0⁻) is negative. (0,0) is neither a max nor a min. f'(3⁻) is negative and f'(3⁺) is positive. Again, (3,−27) is a minimum.

Remember, always substitute into the factored form. When testing, do not evaluate. You are only interested in the sign of the answer.

Possible inflection points f''(x) = 0 = 12x(x − 2). x = 0, x = 2. Substituting back into the original equation (factored) for y, we get the points (0,0) and (2,−16).

Let us use both tests.

TEST 1 f'''(0) and f'''(2) are both not 0. Therefore (0,0) and (2,−16) are inflection points.

TEST 2 f''(0⁻) is positive and f''(0⁺) is negative. Different signs. (0,0) is an inflection point. f''(2⁻) is negative and f''(2⁺) is positive. Different signs. (2,−16) is an inflection point.

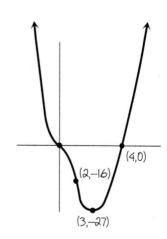

What happens to the ends? For large x,y = (approximately) = x⁴. f(100) is positive and f(−100) is positive. Both ends go to plus infinity.

Now connect the dots.

EXAMPLE 25—

$f(x) = y = x^4 - 12x^2 = x^2(x^2 - 12)$

$\quad = x^2(x - \sqrt{12})(x + \sqrt{12})$

$f'(x) = 4x^3 - 24x = 4x(x^2 - 6)$

$\quad = 4x(x - \sqrt{6})(x + \sqrt{6})$

$f''(x) = 12x^2 - 24 = 12(x^2 - 2)$

$\quad = 12(x - \sqrt{2})(x + \sqrt{2})$

$f'''(x) = 24x$

Intercepts: $y = 0$ $(0,0)$, $(\sqrt{12},0)$, $(-\sqrt{12},0)$. Max, min possibles: $y' = 0$, $x = 0$, $\sqrt{6}$, $-\sqrt{6}$. Substituting for y values in the original equation, we get $(0,0)$, $(\sqrt{6},-36)$, $(-\sqrt{6},-36)$. Testing, $f''(\sqrt{6})$ is positive. $(\sqrt{6},-36)$ is a minimum. $f''(-\sqrt{6})$ is positive. $(-\sqrt{6},-36)$ is a minimum. $f''(0)$ is negative. $(0,0)$ is a maximum.

Possible inflection points: $y'' = 0$. $x = \sqrt{2}$, $-\sqrt{2}$. Substituting for the y value into the original, we get the points $(\sqrt{2},-20)$ and $(-\sqrt{2},-20)$. Testing for inflection points, $f'''(\sqrt{2})$, $f'''(-\sqrt{2}) \neq 0$. Both $(\sqrt{2},-20)$ and $(-\sqrt{2},-20)$ are inflection points.

The ends $f(100)$ and $f(-100)$ are positive. Both ends go to plus infinity. The sketch in two stages is as follows:

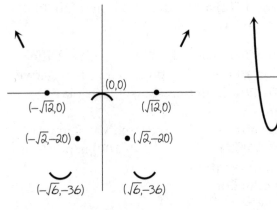

EXAMPLE 26—

$y = f(x) = 9 - x^2$ $-1 \leq x \leq 4$

$y = 9 - x^2 = (3 - x)(3 + x)$

$y' = -2x$

$y'' = -2$

$y''' = 0$

x intercept: $y = 0$ (3,0).

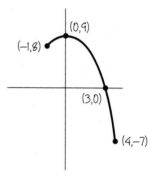

NOTE

$x = -3$ is not in the domain.

   y intercept: $x = 0$ (0,9). Possible max, min: $y' = 0$
$x = 0$. We get the point (0,9). (0,9) is a maximum since
y'' is always negative. No inflection point since y'' is
never equal to 0.

   Since the domain is finite, we must get values for
the left and right ends of $x = -1$, $y = 8$ (−1,8). $x = 4$, $y =$
-7 (4,−7). We see that (4,−7) is an absolute minimum,
(−1,8) is a relative minimum, and (0,9) is an absolute
maximum.

   We will examine several examples with fractional
exponents. We might get cusps or a second kind of
inflection point.

   Given $y = f(x)$

1. $|f'(c)|$ = infinity.

2. $f(c)$ exists [$f(c)$ is some number].

3.   A. $f'(c^-)$ is negative and $f'(c^+)$ is positive, and the
        cusp looks like this:

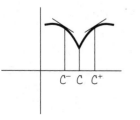

B. f'(c⁻) is positive and f'(c⁺) is negative, and the cusp looks like this:

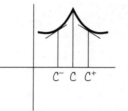

$$c^- \quad c \quad c^+$$

As we will see, if f'(c⁻) and f'(c⁺) have the same sign, we will get another kind of inflection point.

EXAMPLE 27—

$$y = f(x) = (x - 2)^{4/5} + 3$$

$$f'(x) = \left(\frac{4}{5}\right)(x - 2)^{-1/5} = \frac{4}{5(x - 2)^{1/5}}$$

$$f''(x) = \left(\frac{4}{5}\right)\left(-\frac{1}{5}\right)(x - 2)^{-6/5} = \frac{-4}{25(x - 2)^{6/5}}$$

We will test for the cusp first or second kind of inflection point.

1. $f'(2) = \text{infinity}$

2. $f(2) = 3 \ (2,3)$

3. $f'(2^-)$ is negative. $f'(2^+)$ is positive. Cusp with the point down.

x intercept: y = 0. −3 = (x − 2)⁴ᐟ⁵. (x − 2)¹ᐟ⁵ = ±(−3)¹ᐟ⁴, which is imaginary. No x intercepts. y intercept: x = 0. y = (−2)⁴ᐟ⁵ + 3. [0,(−2)⁴ᐟ⁵ + 3] = A. No round max or min and no inflection points. f(1000) is positive and f(−1000) is positive. Both ends go to plus infinity. The sketch:

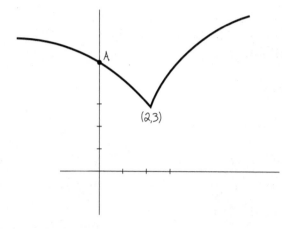

(2,3)

EXAMPLE 28—

$y = f(x) = x^{1/3}$

$$f'(x) = \frac{1}{3x^{2/3}} \qquad f''(x) = \frac{-2}{9x^{5/3}}$$

Intercept (0,0). No round max or min and no inflection points.

1. $|f'(0)| = $ infinity

2. $f(0) = 0$

3. $f'(0^-)$, $f'(0^+)$ both positive. Second kind of inflection point.

4. $f(100,000)$ positive and $f(-100,000)$ negative. The right end goes to plus infinity and the left end goes to minus infinity.

The sketch:

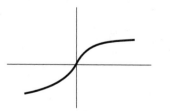

Instructors have been known to give messy examples. While I know yours would never give one, here are two that will take time.

EXAMPLE 29—

$$y = f(x) = \frac{4x}{x^2 + 1}$$

$$y' = 4 \frac{(x^2 + 1)(1) - x(2x)}{(x^2 + 1)^2} = 4 \frac{(1 - x^2)}{(x^2 + 1)^2}$$

$$y'' = 4 \frac{(x^2 + 1)^2(-2x) - (1 - x^2)2(x^2 + 1)(2x)}{(x^2 + 1)^4}$$

$$= 4 \frac{-2x(x^2 + 1)[(x^2 + 1) + 2(1 - x^2)]}{(x^2 + 1)^4} = \frac{-8x(3 - x^2)}{(x^2 + 1)^3}$$

No third derivative . . . too messy!

Intercept (0,0). No vertical asymptote since $x^2 + 1 = 0$ has only imaginary roots. Horizontal asymptote $y = 0$ since the degree of the top is less than the degree of the bottom.

Possible max, min: $y' = 0$ $1 - x^2 = 0$, $x = \pm1$. Substituting in the original, we get (1,2) and (−1,−2). $f''(1)$ is negative. (1,2) is a maximum. $f''(-1)$ is positive. (−1,−2) is a minimum.

Possible inflection points: $y'' = 0$, $-8x(3 - x^2) = 0$. $x = 0, \pm\sqrt{3}$. Substituting into the original, we get the points (0,0), $(\sqrt{3}, \sqrt{3})$, $(-\sqrt{3}, -\sqrt{3})$.

By the messier test, all three are inflection points.

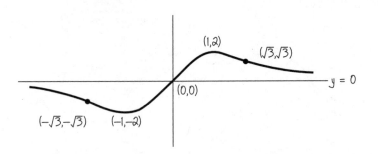

If you liked that one, you'll love this one.

EXAMPLE 30—

$$-1 \le x \le 5 \quad y = x^{2/3}(x^2 - 8x + 16) = x^{2/3}(x-4)^2$$

$$= x^{8/3} - 8x^{5/3} + 16x^{2/3}$$

$$y' = \left(\frac{8}{3}\right)x^{5/3} - \left(\frac{40}{3}\right)x^{2/3} + \left(\frac{32}{3}\right)x^{-1/3}$$

$$= \left(\frac{8}{3}\right)x^{-1/3}(x^2 - 5x + 4) = \frac{8(x-4)(x-1)}{3x^{1/3}}$$

$$y'' = \left(\frac{40}{9}\right)x^{2/3} - \left(\frac{80}{9}\right)x^{-1/3} - \left(\frac{32}{9}\right)x^{-4/3} = \frac{8(5x^2 - 10x - 4)}{9x^{4/3}}$$

There are no asymptotes. Intercepts: $y = 0$ (0,0), (4,0). $y' = 0$, $x = 1,4$. Substituting in the original, we get (1,9), (4,0). $f''(1)$ is negative. (1,9) is a maximum. $f''(4)$ is positive. (4,0) is a minimum.

Possible inflection points: $y'' = 0$. Using the quadratic formula and then substituting these values into the original, we get approximately (2.34,4.86) and (−0.34,9.23). These are both inflection points. Test for cusps: $|f'(0)|$ is infinite. $f(0) = 0$. $f'(0^-)$ is negative and $f'(0^+)$ is positive. Cusp points down. Left-end and right-end substitution: $f(-1) = 25$ (−1,25), $f(5) = 5^{2/3}$. The point is approximately (5,2.92).

Whew!!!! Let us sketch:

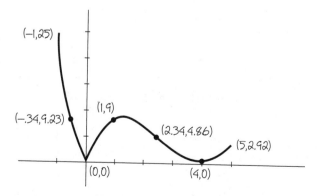

OTHER AIDS

Symmetry:

1. $f(x) = f(-x)$. Symmetrical about the y axis.

2. $f(x) = -f(-x)$. Symmetrical about the origin.

3. $g(y) = g(-y)$. Symmetrical about the x axis. This is only found in graphing curves that are not functions of x.

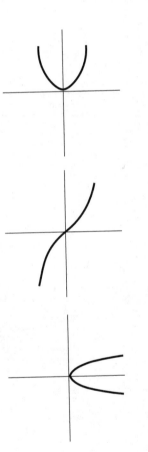

EXAMPLE 31—

y axis symmetry

$f(x) = x^2$

$f(-x) = (-x)^2 = x^2$

EXAMPLE 32—

Symmetric origin

$f(x) = x^3$

$-f(-x) = -(-x)^3 = x^3$

EXAMPLE 33—

x axis symmetry

$g(y) = y^4$

$g(-y) = (-y)^4 = y^4$

For completeness, we will sketch a curve that is not a function of x or y.

EXAMPLE 34—

$$y^2 = \frac{3x^2}{x^2 - 1} \qquad y = \pm\sqrt{\frac{3x^2}{x^2 - 1}}$$

Intercept is (0,0). It turns out to be an isolated point since for all x values between −1 and 1, except x = 0,

$$\frac{x^2}{x^2 - 1}$$

is negative, making y imaginary.

Vertical asymptotes: x = 1, x = −1. Horizontal asymptotes: If we could talk about a degree, the degree of the bottom is 1 and the degree of the top is 1. Horizontal asymptote is y = ±√3. Symmetry with respect to x,y axes. The sketch is as follows:

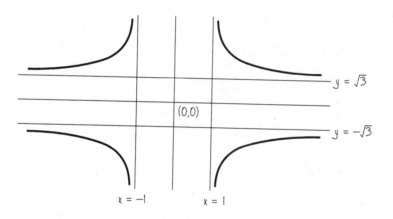

That's about it. You could write a whole book on curve sketching, a large book. I hope this gives you a solid idea on how to sketch curves.

Sometimes, you might like to know about the curve without actually drawing the curve itself. What questions may you ask? Well I'll tell you.

EXAMPLE 35—

For y = x³ − 3x², where is the function increasing and decreasing? What about its concavity?

y increases if the slope of y′ is positive. Look at any graph that increases.

$$y' = 3x^2 - 6x = 3x(x - 2) > 0$$

Solving this quadratic inequality, we find that the curve increases if $x > 2$ or $x < 0$. That would mean y decreases when $0 < x < 2$.

Before, we found that a curve faces up if $y'' > 0$. $y'' = 6x - 6 > 0$ if $x > 1$, which is where the curve is up, and $y'' < 0$ if $x < 1$, where the curve is down.

You might try to do a sketch of this one on your own. If so, don't look at what follows.

Intercepts: $(0,0)$, $(3,0)$. Max, min: $(0,0)$ is the max, $(2,-4)$ is the min. Inflection point: $(1,-2)$. Right end to plus infinity and left end to minus infinity. Polynomials have no asymptotes. And the sketch is . . .

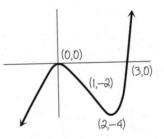

ANSWERS

Answers to the problem on page 68:

x	A	B	C	D	E	F	G	H	I	J	K
f(x)	–	–	+	0	?	–	0	+	+	+	+
f'(x)	+	?	0	–	?	+	+	–	0	?	–
f''(x)	0	?	–	–	?	0	+	+	+	?	0

CHAPTER 4

WORD PROBLEMS MADE EASY . . . WELL, LESS DIFFICULT

Word problems are not difficult because of the calculus. In fact, in more than 90% of word problems, the calculus is very easy. The problem is that few precalculus courses properly prepare you (or prepare you at all) for the setting up of the word problems. As you will see, the algebra is the difficult part of the problems.

The best we can do is to work out a number of word problems that are found in many calculus books. Many of the techniques involved in these problems are applicable to other word problems.

max, min

The theory of these problems is simplicity itself: to find a maximum or a minimum, take the derivative, and set it equal to 0.

In general, there is a picture to be drawn. *Always draw the picture!* Next we have to assign the variable or variables in the problem. Hopefully, by doing enough good examples, you will see how this is done. (I will try my best—you must try your best, which means *don't panic*.) Most of the problems will have two equations in two unknowns. One of these is equal to a number. You will solve for one of the variables and substitute it in the second equation. In the second equation you will take the derivative and set it equal to 0.

Let's start with an easy one.

EXAMPLE 1—

A farmer wishes to make a small rectangular garden with one side against the barn. If he has 200 feet of fence, find the garden of maximum area.

First we make what I call my crummy little picture. You only have to make the picture good enough so you can understand what you drew.

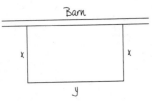

Second, we assign variables. In this case, this task is easy. The two equations involve the area and the perimeter. A = xy. p = 2x + y. The trick, if you could call it a trick, is that the barn is one side of the rectangle. The perimeter involves counting y once.

$$p = 200 = 2x + y \qquad \text{so} \qquad y = 200 - 2x$$

$$A = xy = x(200 - 2x) = 200x - 2x^2$$

$$A' = 200 - 4x = 0 \qquad x = 50 \qquad y = 200 - 2x = 100$$

The area is x times y, which is 5000 square feet. Sometimes the problem only asks for the dimensions, in which case 50 feet and 100 feet are the answers.

To see whether this is truly a max, we find $A''(50)$. Since $A'' = -4$, x = 50 is a maximum. We will check only a few times. Sometimes it is too messy. Usually it just takes more time than I feel like taking.

EXAMPLE 2—

A Texas rancher has an 800-square-mile rectangular plot of land he wishes to divide into three equal regions, as pictured below.

A. Find the dimensions so that the rancher uses the least amount of fencing.

B. If it costs 4 times as much per mile to fence the outside, find the dimensions to minimize the cost.

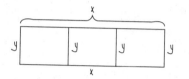

There are two separate problems here. They are done similarly. It is necessary to do them separately, of course.

A. The area in each case equals length times the width. $A = xy = 800$. So $y = 800/x$. The fence length $f = 2x + 4y = 2x + 4(800/x) = 2x + 3200x^{-1}$.

$$f' = 2 - 3200x^{-2} = 2 - \frac{3200}{x^2} = 0$$

$$2x^2 = 3200$$

$$x^2 = 1600$$

$$x = 40$$

$$y = 800/x = 20$$

B. We do not know the cost. So let C equal the cost per mile of the inner fence and 4C the cost per mile of the outer fence. (C is a constant, which we don't know.) The cost of the fence equals the number of fencing miles times the cost per mile.

	Miles of Fence	x	Cost per Mile	=	Total Cost
Outside fence	$2x + 2y$		$4C$		$4C(2x + 2y)$
Inside fence	$2y$		C		$2Cy$

Total cost $T = 4C(2x + 2y) + 2Cy = 8Cx + 10Cy$
But $A = xy = 800$. So y still equals 800/x. Substitute into T.

$T = 8Cx + 10C(800/x) = 8Cx + 8000Cx^{-1}$

$T' = 8C - 8000Cx^{-2} = 8C - \dfrac{8000C}{x^2} = 0$

$8Cx^2 = 8000C$

$x^2 = 1000$

$x = 10\sqrt{10}$ miles

$y = 800/x = 800/10\sqrt{10} = 8\sqrt{10}$ miles

EXAMPLE 3—

An open box with a square bottom is to be cut from a piece of cardboard 10 feet by 10 feet by cutting out the corners and folding the sides up. Find the dimensions that will result in the largest volume.

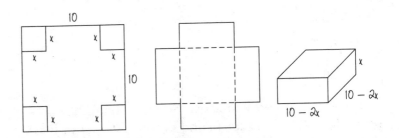

The most difficult part of this problem is the picture, which is given above in three steps. The volume of a box is length times width times height.

$V = (10 - 2x)(10 - 2x)x = 4x^3 - 40x^2 + 100x$

$V' = 12x^2 - 80x + 100$

$\quad = 4(3x^2 - 20x + 25) = 4(3x - 5)(x - 5) = 0$

$x = 5/3$ and 5

We reject 5 since it will give a volume of 0, a minimum.

$$V''(x) = 24x - 80 = 24\left(\frac{5}{3}\right) - 80 = -40$$

indicating a maximum.

The length and width are each $10 - 2x = 10 - 2(5/3) = 20/3$. The box we are looking for is 20/3 by 20/3 by 5/3 feet.

Let's do one more box problem.

EXAMPLE 4—

A box has a square base and no top.

A. Find the minimum surface area needed if the volume is 4 cubic feet. Let's also do the related problem—

B. Find the maximum volume if the surface area is 12 square feet.

As before, the volume = (1)(w)(h). With a square base, $V = x^2y$. A box with no top has five surfaces—a square bottom and four sides, all of which have the same dimensions. The surface area $S = x^2 + 4xy$.

Let's do A.

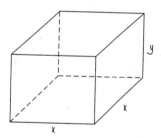

$$V = x^2y = 4 \qquad y = \frac{4}{x^2}$$

$$S = x^2 + 4xy = x^2 + 4x\left(\frac{4}{x^2}\right) = x^2 + 16x^{-1}$$

$$S' = 2x - 16x^{-2} = 2x - \frac{16}{x^2} = 0$$

$$2x^3 = 16 \qquad x^3 = 8 \qquad x = 2 \qquad y = \frac{4}{x^2} = 1$$

The box is 2 by 2 by 1 feet. $S = x^2 + 4xy = (2)^2 + 4(2)(1) = 12$ square feet. $S'' = 2 + 32x^{-3}$. $S''(2) = 2 + 32/8 = 6$, which is positive, indicating a minimum.

Now let's do B. In this case we are maximizing $V = x^2y$ if the surface area $S = x^2 + 4xy = 12$. $y = (12 - x^2)/4x$.

$$V = x^2y = x^2 \frac{12 - x^2}{4x} = \frac{x(12 - x^2)}{4} = 3x - \frac{x^3}{4}$$

$$V' = 3 - \frac{3x^2}{4} = 0 \qquad 3x^2 = 12, \, x^2 = 4, \, x = 2$$

$$y = \frac{12 - x^2}{4x} = 1$$

The box is 2 by 2 by 1 feet, and its maximum volume is 4 cubic feet. Of course, we knew this from the first part of the problem. To check this, $V'' = -3x/2$. $V''(2) = -3$, which is negative, indicating a maximum.

Let's try one that really doesn't have a picture.

EXAMPLE 5

An orchard has 50 apple trees per acre. The average number of apples per tree is 990. For every additional tree per acre planted, each apple tree will give 15 less apples. Find the number of additional trees per acre that should be planted to give the largest number of apples per acre.

The total number of apples equals the product of the number of trees and the number of apples per tree. If we add x trees, we must subtract 15x apples per tree. There are $x + 50$ trees and $990 - 15x$ apples per tree. The total apples $A = (x + 50)(990 - 15x) = 49{,}500 + 240x - 15x^2$. $A' = 240 - 30x = 0$. $x = 8$ trees.

NOTE

This problem is very similar to the minimizing of the cost of the fencing in Example 2, part B (the multiplication part). To become good at these problems, you must notice similarities in problems and use similar techniques in the problems. Then and only then will you become dynamite in these problems. Also note that no matter how good you get, there will always be some problems that will give you trouble.

EXAMPLE 6—

A printer is to use a page of 108 square inches with 1-inch margins at the sides and bottom and a ½-inch margin at the top. What dimensions should the page be so that the area of the printed matter will be a maximum?

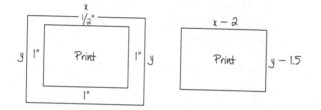

The area of the whole page is $xy = 108$. So $y = 108/x$. The area of the printed matter $A = (x - 2)(y - 1.5) = xy - 1.5x - 2y + 3$. Substituting $y = 108/x$, we get

$$A = 108 - 1.5x - 2\left(\frac{108}{x}\right) + 3$$

$$A = 111 - 1.5x - 216x^{-1}$$

$$A' = -1.5 + 216x^{-2} = -1.5 + \frac{216}{x^2} = 0$$

$$1.5x^2 = 216 \qquad x^2 = 144 \qquad x = 12$$

$$y = \frac{108}{x} \qquad y = 9$$

The page should be 12 by 9 to have the largest amount of print.

EXAMPLE 7—

The strength of a rectangular beam varies jointly as the width and the square of its depth. Which rectangular beam that can be cut from a circular log of radius 10 inches will have maximum strength?

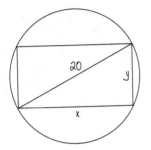

If we let x be the width and y be the depth, we can write the equation without a picture. The strength S = kxy^2; k is an unknown constant. To find a relationship between x and y, we need a picture of the log.

One of the things we always look for is the Pythagorean relationship. In this case $x^2 + y^2 = 400$ (the square of the diameter). In the original equation, it is easier to solve for y^2, because if we solved for x we would have a square root, which would make the derivative much more difficult and sometimes impossible to finish. Therefore

$$y^2 = 400 - x^2 \qquad S = kxy^2 = kx(400 - x^2) = 400kx - kx^3$$

$$S' = 400k - 3kx^2 = 0$$

$$3kx^2 = 400k \qquad x^2 = 400/3$$

$$x = \frac{20\sqrt{3}}{3}$$

$$y^2 = 400 - \frac{400}{3} = \frac{800}{3}$$

$$y = \frac{20\sqrt{6}}{3} \text{ inches}$$

The dimensions to give the maximum strength are

$\frac{20\sqrt{3}}{3}$ inches wide and $\frac{20\sqrt{6}}{3}$ inches deep.

EXAMPLE 8—

Find the shortest distance from $y^2 = 2x$ to the point (2,0).

Although we will draw the picture, it is not truly needed here. Distance means the distance formula. A trick that can always be used is this: instead of using the distance formula (which involves a square root), we can use the square of the distance formula (since the distance is a minimum if the square of the distance is the minimum of the square of all the distances).

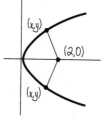

From the picture, we see that two solutions are possible.

Let the square of the distance be $H = (x - 2)^2 + (y - 0)^2 = x^2 - 4x + 4 + y^2$. But on the curve, $y^2 = 2x$. Therefore

$$H = x^2 - 4x + 4 + 2x = x^2 - 2x + 4$$

$$H' = 2x - 2 = 0 \qquad x = 1$$

$$y^2 = 2(1) \qquad \text{so} \qquad y = \pm\sqrt{2}$$

The closest points are $(1, \pm\sqrt{2})$. To find the exact minimum distance, substitute in the distance formula.

The next is a standard-type problem, and many similar kinds are found in all calculus books.

EXAMPLE 9—

Find the dimensions of the largest rectangle (in area) that can be inscribed in the parabola $y = 12 - x^2$, with two vertices of the rectangle on the x axis and two vertices on the curve $y = 12 - x^2$.

Since $y = 12 - x^2$ is symmetric with respect to the y axis, the inscribed rectangle is also symmetric to the y axis. From the picture, the height of the rectangle is y and the base is $x - (-x) = 2x$.

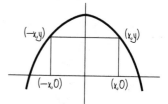

The area $A = 2xy$. Since the point (x,y) is on the curve, $y = 12 - x^2$. Sooooo . . .

$$A = 2xy = 2x(12 - x^2) = 24x - 2x^3$$

$$A' = 24 - 6x^2 = 0 \qquad x = \pm 2$$

only +2 is used—it is a length

$$y = 12 - x^2 = 8$$

The area is $2xy = 2(2)(8) = 32$.

EXAMPLE 10

A rectangle is to be inscribed in a right triangle of sides 6, 8, 10 so that two of the sides are on the triangle. Find the rectangle of maximum area.

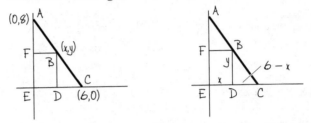

We must note three things in this problem. First, we must set up the triangle in terms of an x-y coordinate system, with the legs on the axes. Second, we note that point B, wherever it is, is represented by the point (x,y), so that the area of the rectangle A = xy. The third thing we must note is that the way to relate x to y is the similar triangles BCD and ACE. Since EC = 6, and ED = x, DC = 6 − x. The proportion we get is

$$\frac{BD}{DC} = \frac{AE}{EC} \quad \text{or} \quad \frac{y}{6-x} = \frac{8}{6}$$

Solving for y, we get

$$y = \frac{8(6-x)}{6} = 8 - \frac{4x}{3}$$

$$A = x\left(8 - \frac{4x}{3}\right) = 8x - \frac{4x^2}{3}$$

$$A' = 8 - \frac{8x}{3} = 0$$

$$x = 3 \quad y = 8 - \frac{4(3)}{3} \quad y = 4$$

The rectangle of maximum area is 4(3) = 12.

EXAMPLE 11

Given a 68-inch string, divide it into two pieces, one a square and one a rectangle whose length is twice its

width, and find the minimum and maximum possible
total area of the figures.

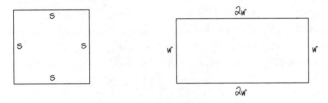

Let one piece be x. The other one will be 68 − x.

If x is the perimeter of the rectangle, then x = 6w. So
w = x/6 annnd ℓ = 2w = x/3. The area of the rectangle
is $x^2/18$.

The perimeter of the square is 68 − x. Since p = 4s, s =
p/4 = (68 − x)/4. The area of the square is $(68 − x)^2/16$.

The total area A = $x^2/18 + (68 − x)^2/16$. dA/dx = $x^2/9$ −
(68 − x)/8 = 0.

Cross multiplying after bringing the negative term to
the other side, we get

8x = 9(68 − x)   8x = 612 − 9x   17x = 612. Sooo x = 36.

The area would be $36^2/18 + (68 − 36)^2/16$ = 72 + 64 = 136.

But there are two other possible values for the max
and min: if the string is uncut, the whole string may
be a rectangle or the whole string might be a square.
If the string is a square, s = p/4 = 68/4 = 17. A = s^2 =
289. If the string is a rectangle, w = p/6 = 68/6; ℓ =
2w = 68/3.

$$A = 1 \times w \qquad w = \frac{68}{3} \text{ times } \frac{68}{6} = 256.9$$

The largest number is the maximum area, when the figure is a square (289). The smallest number is the minimum area, when the string is cut (36″ for the perimeter
of the rectangle, 32″ for the square, minimum area =
136).

So far we have limited the problems to the relatively
gentle. However, there are some pretty nasty problems
in some books. Let's try four of them.

NOTE

In some books, the
string is cut into a
square and an equilat-
eral triangle, or some-
times——yech!——a
square and a circle
(quite messy).

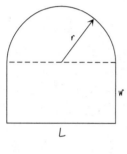

r

w

L

EXAMPLE 12—

Joan's house has a window in the shape of a rectangle surmounted with a semicircle. For a given perimeter p, what are the dimensions of the window if it allows the maximum amount of light?

We first note that maximum light means maximum area. If we are careful and a little clever, the problem is messy but not too bad.

We note we must maximize the area, which is a rectangle plus half a circle.

$$A = Lw + \frac{\pi r^2}{2} = Lw + \frac{\pi (L/2)^2}{2} = Lw + \frac{\pi L^2}{8}$$

The perimeter consists of three sides of the rectangle plus the circumference of the semicircle.

$$p = L + 2w + \frac{2\pi r}{2} = L + 2w + \pi(L/2) = (1 + \pi/2)L + 2w$$

We then solve for w and substitute in the formula for A. We get

$$A = \frac{pL}{2} - (\tfrac{1}{2} + \pi/8)L^2$$

Setting $dA/dL = 0$, we get $L = 2p/(4 + \pi)$. From the formula for p we get $2w = 2p/(4 + \pi)$!!!!!!!!!!!! The maximum light occurs when the length equals twice the width!!!!?!!!

There are many problems where we are asked to find the maximum surface area or volume of a figure inscribed in a sphere or other shape. We will look at a messy one.

EXAMPLE 13—

Find the dimensions of the cone of largest volume that can be inscribed in a sphere of radius R.

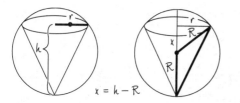

First we note the symmetry of the cone inside the sphere. We note that the variables are r and h, the dimensions of the cone, which are to be found. The only known in the problem is the radius of the sphere R, which is a given number. In the second picture is the way h, r, and R are related, again a right triangle.

$$(h - R)^2 + r^2 = R^2 \qquad \text{or} \qquad r^2 = 2hR - h^2$$

Since the volume of a cone is $(1/3)\pi r^2 h$, it will make a much easier problem to differentiate if we solve for r^2. Thus

$$V = \frac{\pi h(2hR - h^2)}{3} = \frac{\pi(2h^2R - h^3)}{3}$$

$$V' = (\pi/3)(4hR - 3h^2) = 0$$

$$3h^2 = 4hR \qquad h = 4R/3$$

We can then substitute into the expression for r^2, and then take the square root to get r.

At this point you might cry, "No more. *No more!!!!*" However, there are two other problems which are in most books. I think you will appreciate seeing them.

EXAMPLE 14—

A fence is 8 feet tall and is on level ground parallel to a building. Find the shortest ladder that will go from the ground over the fence to the building if the fence is 1 foot from the wall.

We will do this problem two different ways; neither will delight you, but both are very instructive.

The ladder L is AE. BC = 1. Again we note similar triangles ABD, ACE. We will let AB = x. By the

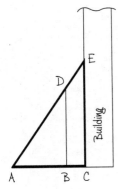

Pythagorean theorem, $AD = (x^2 + 64)^{1/2}$. By similar triangles,

$$\frac{AB}{AD} = \frac{AC}{AE} \quad \text{or} \quad \frac{x}{(x^2 + 64)^{1/2}} = \frac{x + 1}{L}$$

$$L = \frac{(x + 1)(x^2 + 64)^{1/2}}{x} = (x^2 + 64)^{1/2} + \frac{(x^2 + 64)^{1/2}}{x}$$

$$0 = L' = \frac{x}{(x^2 + 64)^{1/2}} + \frac{\dfrac{x^2}{(x^2 + 64)^{1/2}} - (x^2 + 64)^{1/2}}{x^2}$$

$$L' = \frac{x}{(x^2 + 64)^{1/2}} + \frac{1}{(x^2 + 64)^{1/2}} - \frac{(x^2 + 64)^{1/2}}{x^2} = 0$$

$$\frac{x + 1}{(x^2 + 64)^{1/2}} = \frac{(x^2 + 64)^{1/2}}{x^2}$$

$$x^2(x + 1) = [(x^2 + 64)^{1/2}]^2$$

$$x^3 + x^2 = x^2 + 64 \qquad x^3 = 64 \qquad x = 4 \text{ feet!!!}$$

$$L = \frac{(x + 1)(x^2 + 64)^{1/2}}{x} = \frac{(4 + 1)(16 + 64)^{1/2}}{4} = \frac{5(80)^{1/2}}{4}$$

$$= 5\sqrt{5} \text{ feet}$$

Yeeeow!!!

The second solution is not much nicer. We will do this with trig functions in terms of the angle A.

cot A = AB/8 (always try to get the unknown on top). AB = 8 cot A. AC = 8 cot A + 1.

sec A = AE/AC. AE = L. L = AC sec A = (8 cot A + 1) sec A = 8 csc A + sec A (using trig identities).

$$0 = \frac{dL}{dA} = -8 \cot A \csc A + \tan A \sec A$$

$$8 \cot A \csc A = \tan A \sec A$$

$$8 \,\frac{\cos A}{\sin A} \cdot \frac{1}{\sin A} = \frac{\sin A}{\cos A} \cdot \frac{1}{\cos A}$$

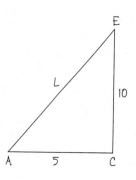

$$\frac{\sin^3 A}{\cos^3 A} = 8 \qquad \tan^3 A = 8 \qquad \tan A = 2$$

This says the ratio of DB to AB is 2:1. DB = 8 feet; AB = 4 feet. Therefore AC = 4 + 1 = 5. By the same reasoning, EC must be 10 feet.

$$L = (5^2 + 10^2)^{1/2} = (125)^{1/2} = 5\sqrt{5} \text{ feet!!!}$$

We finish the max and min problems with what I consider the ultimate in distance problems—not the hardest, but one of the slickest.

EXAMPLE 15—

A man is on an island which is 4 miles from the nearest point on a straight shoreline. He wishes to go to a house which is 12 miles from this nearest point. If he rows at 3 miles per hour and runs at 5 miles per hour, find the shortest time to reach the house.

Insight: If he rowed faster than he ran, he would have rowed straight for the house. If he ran muuuch faster than he rowed, he would row straight for the shore and then run. So we are looking for the point on the shore where he must land. The picture is below.

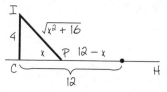

The unknown landing point P is x miles from C, the closest point. We get IP by the Pythagorean theorem. Since CH is 12, PH is 12 − x. We need to set up an equation for time, where time = distance/speed.

	Distance	Speed	Time = $\dfrac{\text{Distance}}{\text{Speed}}$
from I to P (water)	$(x^2 + 16)^{1/2}$	3	$(x^2 + 16)^{1/2}/3$
from P to H (land)	12 − x	5	$(12 - x)/5$

$$\text{Total time } t = \frac{(x^2 + 16)^{1/2}}{3} + \frac{12}{5} - \frac{x}{5}.$$

$$dt/dx = \frac{x}{3(x^2 + 16)^{1/2}} - \frac{1}{5} = 0$$

$$5x = 3(x^2 + 16)^{1/2} \quad 25x^2 = 9(x^2 + 16) \quad 25x^2 = 9x^2 + 144$$

$$16x^2 = 144 \quad x^2 = 9 \quad x = 3$$

$$t = \frac{(3^2 + 16)^{1/2}}{3} + \frac{12 - 3}{5} = \frac{52}{15} = 3 \text{ hours 28 minutes!}$$

Let us now tackle other kinds of word problems, which should be a little easier.

RELATED RATES

The rates we will talk about refer to how things change with respect to time. Related rates are how the rates of two or more variables are connected with each other.

Before we try some problems, we will give an equation and differentiate it with respect to time. Suppose we are given

EXAMPLE 16—

$$z = x^2 + y^3$$

x, y, and z are variables and all are functions of time. What we will do is differentiate the equation implicitly with respect to time. How does this differ from before?

When we were given $y = (x^2 + 1)^{100}$, we let $u = x^2 + 1$, and

$$\frac{dy}{dx} = \frac{dy}{du} \cdot \frac{du}{dx} \qquad \frac{dy}{du} = 100u^{99} \qquad \frac{du}{dx} = 2x$$

so

$$\frac{dy}{dx} = 100u^{99} \cdot 2x = 200(x^2 + 1)^{99}x$$

But if $y = u^{100}$ and we did *not* know what u was (only a function of x),

$$\frac{dy}{dx} = 100u^{99} \cdot \frac{du}{dx}$$

And if $y = u^{100}$ and u was a function of t (which we did not know),

$$\frac{dy}{dt} = 100u^{99} \frac{du}{dt}$$

That's all there is to it.

Let's get back to $z = x^2 + y^3$. Differentiating with respect to time, we get

$$\frac{dz}{dt} = 2x \frac{dx}{dt} + 3y^2 \frac{dy}{dt}$$

How many variables are in this equation? Five!! x, y, dz/dt, dx/dt, and dy/dt. In order to solve this equation, four of the variables must be given or found from the problem.

Let's try one more.

EXAMPLE 17—

$$x^4 y^7 + \tan x = v^5$$

$$x^4 \cdot 7y^6 \frac{dy}{dt} + y^7 \cdot 4x^3 \frac{dx}{dt} + \sec^2 x \cdot \frac{dx}{dt} = 5v^4 \cdot \frac{dv}{dt}$$

Now we are ready for related-rates problems. In general these problems are easier than max and min problems since the most difficult part of the problem, the original equation, is either given or much easier to find than before.

EXAMPLE 18—

The radius of a circle is increasing at 2 feet per minute. Find the rate at which the area is increasing when the radius is 10 feet.

We are given that dr/dt = 2 and we are looking for dA/dt. All we need is to know the area of a circle, differentiate it with respect to time, and substitute to get the answer.

$A = \pi r^2$ $dA/dt = 2\pi r\, dr/dt = 2\pi(10)(2) = 40\pi$ square feet per minute.

Are all the problems this easy? Well, no, but they are not too much more difficult.

EXAMPLE 19—

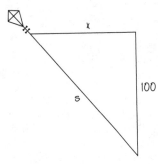

A girl flying a kite plays out a string at 2 feet per second. The kite moves horizontally at an altitude of 100 feet. If there is no sag in the string, find the rate at which the kite is moving when 260 feet of string have been played out.

We first have to imagine a right triangle with the vertical distance, the horizontal distance, and the string. We then must determine which are the constants and which are the variables. The only constant is that the kite is always 100 feet high. (I wonder how they keep the kite exactly 100 feet high?)

Although we are given the length of the string, it is only at a particular instant. The string length, s, and horizontal length, x, are changing.

The equation is $x^2 + 100^2 = s^2$. Differentiating with respect to time, we get

$$2x\,\frac{dx}{dt} = 2s\,\frac{ds}{dt} \qquad \text{or} \qquad x\,\frac{dx}{dt} = s\,\frac{ds}{dt}$$

ds/dt is 2. Since s = 260, the Pythagorean theorem tells us that x = 240 feet. (You should have memorized the common Pythagorean triples. This is a 5,12,13 right tri-

angle, or more precisely, a 10,24,26 right triangle.)
Substituting in

$$x \frac{dx}{dt} = s \frac{ds}{dt}$$

we get

$$(240) \frac{dx}{dt} = (260)(2).$$

$$\frac{dx}{dt} = \frac{13}{6} \text{ feet per second.}$$

EXAMPLE 20—

Sand is leaking from a bag and is forming a cone in
which the radius is 6 times as large as the height. Find
the rate at which the volume is increasing when the
radius is 3 inches and the height is increasing at 2
inches per minute.

We know that the volume of the cone is $V = (1/3)\pi r^2 h$. There are two variables, but we also know
that $r = 6h$. We can write V in terms of one unknown,
but which unknown? Since $dh/dt = 2$ is given, let us
write everything in terms of h.

$$V = (1/3)\pi(6h)^2 h = 12\pi h^3$$

$$\frac{dV}{dt} = 36\pi h^2 \left(\frac{dh}{dt} \right) \qquad r = 6h \text{ and } r = 3; h = \frac{1}{2}$$

$$\frac{dV}{dt} = 36\pi(\tfrac{1}{2})^2(2) = 18\pi \text{ cubic inches per minute}$$

EXAMPLE 21—

When a gas is compressed adiabatically (with no gain
or loss of heat), it satisfies the formula $PV^{1.4} = k$; k is a
constant. Find the rate at which the pressure is changing if the pressure P is 560 pounds per square inch, the

volume is 70 cubic inches, and the volume is increasing at 2 cubic inches per minute.

$PV^{1.4} = k$. Differentiating with respect to time, we get $P(1.4V^{.4}\, dV/dt) + V^{1.4}\, dP/dt = 0$. It looks messy at this point, but fear not.

$$dP/dt = -\frac{1.4PV^{.4}(dV/dt)}{V^{1.4}} = \frac{-1.4P(dV/dt)}{V}$$

Substituting, we get

$$dP/dt = \frac{-(1.4)(560)(2)}{70} = -22.4 \text{ pounds per square inch per minute}$$

EXAMPLE 22—

At 7 a.m. a plane flies over the city going east at 600 mph. At 9 a.m. a plane flies over the city going north at 350 mph. At what rate were the planes separating at 11 a.m.?

The key is to first make a drawing at the middle time, 9 a.m. The first plane is going east at 600 mph × 2 hours = 1200 miles east of the city and keeps on going east. The distance is 1200 + x. At 9 a.m., the other plane is over the city and is going north, distance 0 + y = y. Let z be the distance between the planes.

$$z^2 = (x + 1200)^2 + y^2$$

Find d/dt.

$$2z\frac{dz}{dt} = 2(x + 1200)\frac{dx}{dt} + 2y\frac{dy}{dt}$$

The separating speed

$$\frac{dz}{dt} = \frac{(x + 1200)dx/dt + y\, dy/dt}{z}$$

dx/dt = 600; 9 a.m. to 11 a.m. = 2 hours × 600 = 1200 = x.

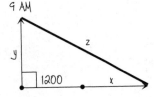

9 AM

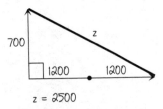

11 AM

z = 2500

$dy/dt = 350$; time = 2 hours $\times$ 350 = 700 = y.

Sticking numbers in the 11 a.m. picture and using old Pythagorus, we get z = 2500 (7,24,25 right triangle).

$$\frac{dz}{dt} = \frac{(1200 + 1200)(600) + (700)(350)}{2500}$$

= 674 mph, the separating speed

Finally, sometimes you need two different related rates equations to solve the problem.

EXAMPLE 23—

Find the rate at which the volume changes with respect to time when the change of surface area with respect to time is 600 square inches per hour and the volume is 1000 cubic inches.

Now $V = x^3$. So $dV/dt = 3x^2\, dx/dt$. We need to know both x and dx/dt, but we are given neither directly. However, $V = x^3 = 1000$. Sooo x = 10 inches. We are given dS/dt. So the surface area of a cube $S = 6x^2$.

$$\frac{dS}{dt} = 12x\,\frac{dx}{dt} \qquad \frac{dS}{dt} = 600 \qquad x = 10$$

$$\frac{dx}{dt} = 5 \text{ inches per hour.}$$

Finally . . .

$$\frac{dV}{dt} = 3x^2\,\frac{dx}{dt} = 3(10)^2\,(5) = 1500 \text{ cubic inches per hour.}$$

THE GRAVITY OF THE SITUATION

The last type of word problem we will deal with is throwing an object into the air. There are three variables that determine how high an object goes. One is gravity. On this planet, gravity is −32 feet per second squared (or −9.8 meters per second squared); the minus sign indicates down. The second is the initial velocity (positive if the object is thrown upwards, negative if the object is thrown down, and zero if the object is dropped). The third is the initial height (it goes higher if I throw something from the Empire State Building than from the ground). Interestingly, the weight has no bearing. If I give a rock or a refrigerator the same initial velocity (which I can't, of course . . . or can I?), both will go just as high. Friction, winds, etc., are not included.

The following symbols are usually used:

v_0: initial velocity—the velocity at time $t = 0$

y_0: initial height—the height at time $t = 0$

The initial height is either zero, the ground, or positive. It is very difficult to throw something upward from under the ground. The acceleration is dv/dt. So the velocity is the antiderivative of the acceleration.

The velocity is dy/dt. The height y is the antiderivative of the velocity.

EXAMPLE 24—

On the planet Calculi, gravity is 20 feet per second squared. A ball is thrown upward—initial velocity 40 feet per second, initial height 50 feet.

A. Write the equation for the height.

B. Find the ball's maximum height.

C. Find the velocity when the ball hits the ground.

A. $v' = a = -20$ (gravity is down). Take antiderivative—

$$v = -20t + C_1. \quad v(0) = 40, \text{ sooooo}$$

$$y' = v = -20t + 40$$

$$y = -10t^2 + 40t + C_2$$

$y(0) = 50$, sooooo the equation for the height is

$$y = -10t^2 + 40t + 50.$$

B. Throw something lightish into the air straight up. You see that the object stops at its maximum height. This means $v = -20t + 40 = 0$. $t = 2$. Substituting into the equation for the height y, we get $y = -10(2)^2 + 40(2) + 50 = 90$ feet, its maximum height.

C. The ground, as noted before, is $y = 0$. So

$$0 = -10t^2 + 40t + 50 = -10(t^2 - 4t - 5)$$

$$= -10(t - 5)(t + 1)$$

Rejecting $t = -1$ (negative time is before the object is thrown upward), we get $t = 5$. Substituting $t = 5$ into the velocity equation, we get . . .

$$v = -20(5) + 40 = -60 \text{ feet per second}$$

where the minus sign indicates that the object is going down when it hits the ground.

NOTE

Since we live on the planet Earth, it would be a good idea to memorize the formulas for throwing an object upward: $y = -16t^2 + v_0t + y_0$ in feet and $y = -4.9t^2 + v_0t + y_0$ in meters.

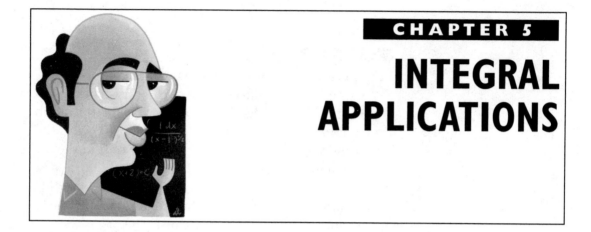

CHAPTER 5

INTEGRAL APPLICATIONS

AREAS

We would like to explore some applications of the integral. The first is the area between two curves. Suppose we have two functions f(x) and g(x), where f(x) is always greater than or equal to g(x). Its picture might look like this:

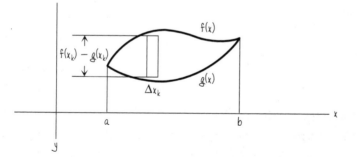

We are going to add these rectangles up. One rectangle is represented by the height times the base = $[f(x_k) - g(x_k)]\Delta x_k$. If we add them up and take the limits properly, we get

$$\int_a^b [f(x) - g(x)]\, dx$$

EXAMPLE I—

Find the area between $y = x + 1$ and $y = 3 - x^2$.

We first draw the curves to see which is the top curve and which is the bottom curve. We next find the limits of integration, the left- and rightmost x values, by setting the curves equal to each other to find the points where they meet.

$x + 1 = 3 - x^2$, $x^2 + x - 2 = 0$, $(x + 2)(x - 1) = 0$. So $x = -2$ and 1. We then set up the integral top-curve-minus-bottom-curve dx from -2 to 1.

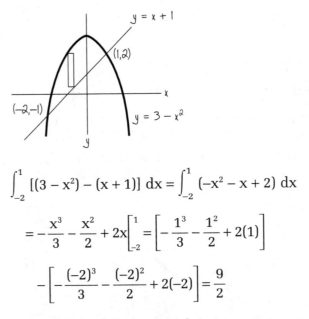

$$\int_{-2}^{1} [(3 - x^2) - (x + 1)]\, dx = \int_{-2}^{1} (-x^2 - x + 2)\, dx$$

$$= -\frac{x^3}{3} - \frac{x^2}{2} + 2x \Big[_{-2}^{1} = \left[-\frac{1^3}{3} - \frac{1^2}{2} + 2(1) \right]$$

$$- \left[-\frac{(-2)^3}{3} - \frac{(-2)^2}{2} + 2(-2) \right] = \frac{9}{2}$$

Sometimes it becomes advantageous to set up the rectangles horizontally. We have h(y) and j(y), where j(y) is always larger than or equal to h(y). The picture would look like this:

The area of each rectangle would again be the base times the height = $[j(y_k) - h(y_k)]\, \Delta y_k$. If we again properly take the limits, we get

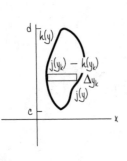

$$\int_{c}^{d} [j(y) - h(y)]\, dy$$

EXAMPLE 2—

Find the area between the curves $x = y^2$ and $x = 3y + 4$.

If we try to make the rectangles as before, the top curve is the parabola, which is OK, but if you look very closely there are two parts of the bottom curve, a little part of the bottom of the parabola and the straight line. We would have to split the region and set up two integrals to find the area. However, the right curve is always the straight line and the left curve is always the parabola. We take right-curve-minus-left-curve dy. To find the lowest and highest y values, we again set the curves equal to each other. $y^2 = 3y + 4$, $y^2 - 3y - 4 = 0$, $(y + 1)(y - 4) = 0$. $y = -1$ and $y = 4$.

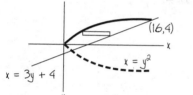

$$\int_{-1}^{4} [(3y + 4) - y^2] \, dy = \frac{3y^2}{2} + 4y - \frac{y^3}{3} \Bigg|_{-1}^{4}$$

$$= \left[\frac{3(4)^2}{2} + 4(4) - \frac{4^3}{3} \right] - \left[\frac{3(-1)^2}{2} + 4(-1) - \frac{(-1)^3}{3} \right] = \frac{125}{6}$$

Sometimes splitting the region cannot be avoided. We will do a problem that must be split no matter which way you draw the rectangles. We will do it two ways. With experience you may be able to tell which way is better before you begin the problem.

EXAMPLE 3—

Find the area of the region bounded by $y = x$ to the point $(2,2)$, then the curve $y = 6 - x^2$ from $(2,2)$ to $(4,-10)$, and then $y = -5x/2$ from $(4,-10)$ to $(0,0)$ where it meets the curve $y = x$.

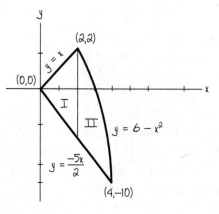

REGION I

$$\int_0^2 \left[x - \left(\frac{-5x}{2} \right) \right] dx = \int_0^2 \frac{7x}{2}\, dx = \frac{7x^2}{4} \Big[_0^2 = \frac{7(2)^2}{4} - 0 = 7$$

REGION II

$$\int_2^4 \left[(6 - x^2) - \left(-\frac{5x}{2} \right) \right] dx = 6x - \frac{x^3}{3} + \frac{5x^2}{4} \Big[_2^4$$

$$= 6(4) - \frac{4^3}{3} + \frac{5(4)^2}{4} - \left[6(2) - \frac{2^3}{3} + \frac{5(2)^2}{4} \right] = 8\tfrac{1}{3}$$

Total area is $7 + 8\tfrac{1}{3} = 15\tfrac{1}{3}$.

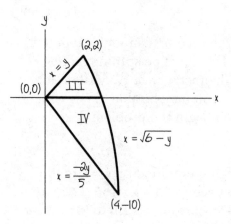

REGION III

$$\int_0^2 [(6-y)^{1/2} - y]\, dy = -\frac{2}{3}(6-y)^{3/2} - \frac{y^2}{2} \Bigg|_0^2$$

$$= \left[-\frac{2}{3}(6-2)^{3/2} - \frac{2^2}{2} \right] - \left[-\frac{2}{3}(6-0)^{3/2} - 0 \right]$$

$$= \frac{2}{3} 6^{3/2} - \frac{22}{3}$$

REGION IV

$$\int_{-10}^0 [(6-y)^{1/2} - (-2/5)y]\, dy = -\frac{2}{3}(6-y)^{3/2} + \frac{y^2}{5} \Bigg|_{-10}^0$$

$$= \left\{ -\frac{2}{3}(6-0)^{3/2} + 0 \right\} - \left\{ -\frac{2}{3}[6-(-10)]^{3/2} + \frac{(-10)^2}{5} \right\}$$

$$= \frac{68}{3} - \frac{2}{3} 6^{3/2}$$

Total area $= \dfrac{2}{3} 6^{3/2} - \dfrac{22}{3} + \dfrac{68}{3} - \dfrac{2}{3} 6^{3/2} = \dfrac{46}{3} = 15\frac{1}{3}.$

The two answers agree. The area is 15⅓.

The last example for us is a region where the curves cross each other (the top curve becomes the bottom and the bottom becomes the top). This is another reason you must draw the region.

EXAMPLE 4—

Find the area of the region between $y = x$ and $y = x^3 - 3x$.

To find the limits, we set $x^3 - 3x = x$. $x^3 - 4x = 0$. $x(x+2) \times (x-2) = 0$. So $x = -2$, 0, and 2. In region I, the top curve is the cubic. So we get $[(x^3 - 3x) - (x)]\, dx$ from -2 to 0. In region II, we have the straight line as the top curve. We get $[x - (x^3 - 3x)]\, dx$ from 0 to 2.

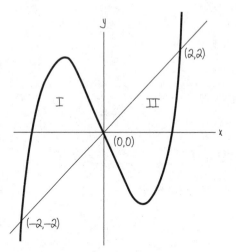

REGION I

$$\int_{-2}^{0} (x^3 - 4x)\ dx = \frac{x^4}{4} - 2x^2 \left[\begin{matrix} 0 \\ -2 \end{matrix} \right. = 0 - \left(\frac{(-2)^4}{4} - 2(-2)^2 \right) = 4$$

REGION II

$$\int_{0}^{2} (4x - x^3)\ dx = 2x^2 - \frac{x^4}{4} \left[\begin{matrix} 2 \\ 0 \end{matrix} \right. = \left[2(2)^2 - \frac{2^4}{4} \right] - 0 = 4$$

The total area is $4 + 4 = 8$.

VOLUMES OF ROTATIONS

The next topic is finding volumes of rotations. This is very visual. If you see the picture, the volume is easy. If not, this topic is very hard.

Imagine a perfectly formed apple with a line through the middle from top to bottom. We can find the volume two different ways. One way is by making slices perpendicular to the line (axis). (We will do the other way later with an onion.) Each slice is a disc, a thin cylinder. Its volume is $\pi r^2 h$, where h is very small. If we add up all the discs, taking the limits properly, we get the volume.

NOTE

If we noticed that the two regions have the same area, we could have found the area of either one and doubled its value.

We will take the same region in six different problems, rotating this region differently six times and getting six different volumes.

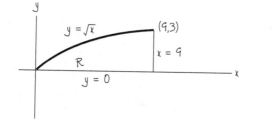

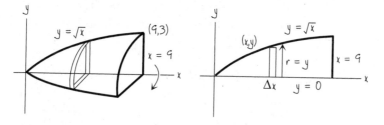

EXAMPLE 5—

Find the volume if the region R is rotated about the x axis.

The volume of each disc is $\pi r^2 h$. $h = \Delta x$. $r = y$. So $r^2 = y^2 = x$; x goes from 0 to 9.

$$V = \int_0^9 \pi r^2\, dh = \pi \int_{x=0}^9 x\, dx = \frac{\pi x^2}{2}\Big[_0^9 = \frac{81\pi}{2}$$

The integrals are almost always easy. Once you understand the picture, all will be easy. But it takes most people time to study the pictures.

Let's get back to our apple. Suppose we core our apple. When we take slices perpendicular to the axis, we get rings. The area of a ring is the area of the outside minus the area of the inside. The volume of each disc is

$$(\pi r^2_{outside} - \pi r^2_{inside})h$$

Again, h is small.

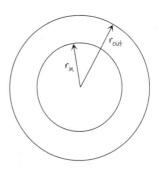

EXAMPLE 6—

Find the volume if our region is rotated about the y axis.

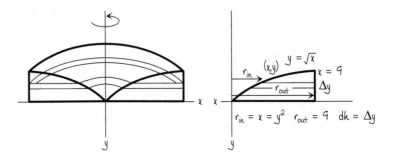

As you rotate this region there is a hole. The outside radius is always 9 and the inside radius is always the x value. But $x = y^2$. $r^2 = x^2 = y^4$.

$$V = \int_{y=0}^{3} (\pi r_{out}^2 - \pi r_{in}^2)\, dh = \pi \int_{0}^{3} (9^2 - y^4)\, dy$$

$$= \pi\left[81y - \frac{y^5}{5} \right]_0^3 = \pi\left[81(3) - \frac{3^5}{5} \right] = \frac{972}{5}\pi$$

EXAMPLE 7—

Find the volume of our glorious region if it is rotated about the line x = 9.

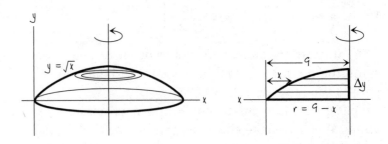

Notice that when we rotate the region about x = 9, there is no hole. $V_{sect} = \pi r^2 h$. $r = 9 - x = 9 - y^2$. $r^2 = 81 - 18y^2 + y^4$. $h = \Delta y$.

$$V = \pi \int_0^3 (81 - 18y^2 + y^4)\, dy = 81y - 6y^3 + \frac{y^5}{5} \Bigg]_0^3 \pi$$

$$= \pi \left[81(3) - 6(3)^3 + \frac{3^5}{5} - 0 \right] = \frac{648\pi}{5}$$

EXAMPLE 8—

Find the volume if the same region R is rotated about the line $x = -1$.

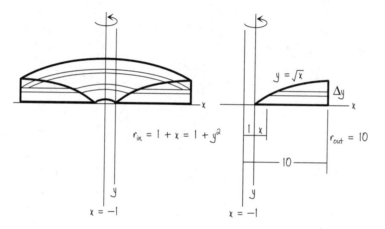

$V_{section} = \pi(r_{out}^2 - r_{in}^2)h.$ $r_{out} = 9 + 1 = 10.$ $r_{in} = 1 + x =$
$1 + y^2.$ $r^2 = 1 + 2y^2 + y^4.$ $h = \Delta y.$

$$V = \int_0^3 (\pi r_{out}^2 - \pi r_{in}^2)\, dy = \pi \int_0^3 [10^2 - (1 + 2y^2 + y^4)]\, dy$$

$$= \pi \int_0^3 (99 - 2y^2 - y^4)\, dy = 99y - \frac{2y^3}{3} - \frac{y^5}{5} \Bigg]_0^3 \pi$$

$$= \pi \left[99(3) - \frac{2(3)^3}{3} - \frac{3^5}{5} - 0 \right] = \frac{1152\pi}{5}$$

EXAMPLE 9—

Find the volume if our beloved region is rotated about $y = 5$.

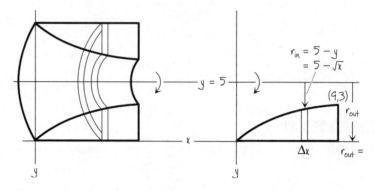

$V_{\text{section}} = \pi(r_{\text{out}}^2 - r_{\text{in}}^2)h$. $r_{\text{out}} = 5$. $r_{\text{in}} = 5 - y = 5 - x^{1/2}$. $r_{\text{in}}^2 = 25 - 10x^{1/2} + x$. $h = \Delta x$.

$$V = \int_0^9 (\pi r_{\text{out}}^2 - \pi r_{\text{in}}^2)\, dx = \pi \int_0^9 [25 - (25 - 10x^{1/2} + x)]\, dx$$

$$= \pi \int_0^9 (10x^{1/2} - x)\, dx = \frac{20}{3} x^{3/2} - \frac{x^2}{2} \bigg|_0^9 \pi$$

$$= \pi \left(\frac{20}{3} 9^{3/2} - \frac{9^2}{2} - 0 \right) = 139.5\pi$$

EXAMPLE 10—

And for our final attraction we will take the same region and rotate it about the line $y = -2$.

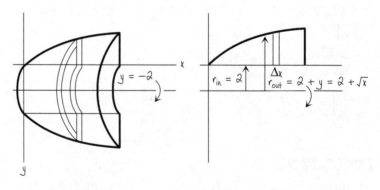

$V_{\text{sect}} = (r_{\text{out}}^2 - r_{\text{in}}^2)h$. $r_{\text{in}} = 2$, $r_{\text{out}} = 2 + y = 2 + x^{1/2}$. $r_{\text{out}}^2 = 4 + 4x^{1/2} + x$.

$$V = \int_0^9 (\pi r_{\text{out}}^2 - \pi r_{\text{in}}^2)\, dx = \pi \int_0^9 [(4 + 4x^{1/2} + x) - 4]\, dx$$

$$= \pi \int_0^9 (4x^{1/2} + x)\, dx = \frac{8}{3} x^{3/2} + \frac{x^2}{2} \Bigg[_0^9 \pi$$

$$= \pi \left(\frac{8}{3} \cdot 9^{3/2} + \frac{9^2}{2} \right) = \frac{225\pi}{2}$$

The next kind of volumes we will consider are rotations again, but we will do it a different way. Think about an onion with each layer a whole piece. We will add up layer by layer until we get a volume. We will add up cylindrical shells, tall cylindrical shells. We will see one of these shells, a general picture, and two examples.

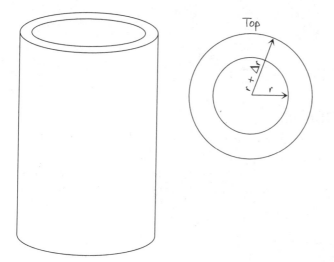

$$V_{\text{shell}} = \pi r_{\text{out}}^2 h - \pi r_{\text{in}}^2 h$$

$$= \pi h (r_{\text{out}}^2 - r_{\text{in}}^2)$$

$$= \pi h (r_{\text{out}} + r_{\text{in}})(r_{\text{out}} - r_{\text{in}})$$

$$= 2\pi h \left(\frac{r_{\text{out}} + r_{\text{in}}}{2} \right)(\Delta r)$$

$$= 2\pi \cdot \text{height} \cdot \text{average radius} \cdot \text{thickness}$$

This is an example of a rotation about the y axis of the region bounded by y = f(x), y = 0, x = a, x = b. *Notice that the axis of the cylinder is the axis which the curve is rotated around.*

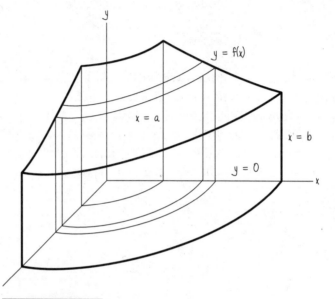

EXAMPLE 11—

Let us try to do the same region R we had before and rotate it about the y axis. We take the same region for two reasons: (1) you do not have to worry about different curves, and (2) we would like to show that the answers are the same.

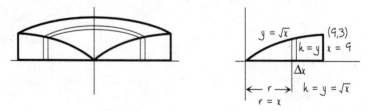

$$V = \int 2\pi \cdot \text{average radius} \cdot \text{average height} \cdot \text{thickness}$$

$$= 2\pi \int_0^9 x \cdot x^{1/2} \, dx = 2\pi \int_0^9 x^{3/2} \, dx = \frac{4}{5} x^{5/2} \left. \right]_0^9 \pi$$

$$= \frac{4\pi}{5} (9^{5/2} - 0) = \frac{972\pi}{5}$$

EXAMPLE 12—

We will do our region one last time about the x axis.

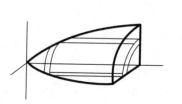 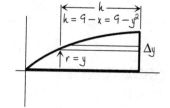

$$V = 2\pi \int_{y=0}^{3} \text{average radius} \cdot \text{average height} \cdot \text{thickness}$$

$$= 2\pi \int_{0}^{3} y(9 - y^2) \, dy = 2\pi \int_{0}^{3} (9y - y^3) \, dy$$

$$= \frac{9y^2}{2} - \frac{y^4}{4} \left[\begin{array}{c} 3 \\ \\ 0 \end{array}\right. 2\pi = 2\pi\left[\frac{9(3)^2}{2} - \frac{3^4}{2}\right] = \frac{81\pi}{2}$$

VOLUMES BY SECTION

The last part of this chapter is volumes by section. The sections are not circles but other shapes. Two examples are given.

EXAMPLE 13—

Find the volume of the following figure: Base bounded by $y = \frac{1}{4}x^2$ and $y = 4$ in the x-y plane. Sections perpendicular to the y axis are rectangles with heights of $\frac{1}{3}$ the base.

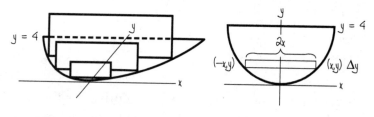

$$A = bh = b(\tfrac{1}{3}b) = b^2/3 \qquad b = x - (-x) = 2x$$

$$y = \tfrac{1}{4}x^2 \qquad x^2 = 4y \qquad x = 2y^{1/2} \qquad 2x = 4y^{1/2} = b$$

$A = (4y^{1/2})^2/3 = 16y/3$

Volume of each section $= (16y/3)\Delta y$

$$V = \int_0^4 \frac{16y}{3} \, dy = \frac{8y^2}{3} \Big|_0^4 = \frac{8(4)^2}{3} - \frac{8(0)^2}{3} = \frac{128}{3}$$

EXAMPLE 14—

Find the volume of the following figure: Again the base is bounded by $y = \frac{1}{4}x^2$ and $y = 4$ in the x-y plane. Sections perpendicular to the x axis are right isosceles triangles with one of the legs on the x-y plane.

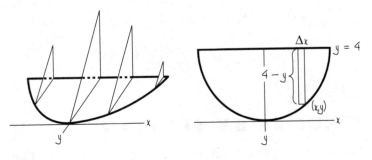

$A = \frac{1}{2}bh = \frac{1}{2}b^2$ (*isosceles* means two equal legs)

$$b = 4 - y = 4 - \frac{x^2}{4}$$

$$A = \frac{1}{2}\left(4 - \frac{x^2}{4}\right)^2 = 8 - x^2 + \frac{x^4}{32}$$

Volume of each section $= (8 - x^2 + x^4/32)\Delta x$

$$V = \int_{-2}^{2} \left(8 - x^2 + \frac{x^4}{32}\right) dx = 8x - \frac{x^3}{3} + \frac{x^5}{160} \Big|_{-2}^{2}$$

$$= \left[8(2) - \frac{2^3}{3} + \frac{2^5}{32}\right] - \left[8(-2) - \frac{(-2)^3}{3} + \frac{(-2)^5}{32}\right] = \frac{406}{15}$$

If you draw your pictures very carefully, this figure is symmetric with respect to the y axis. You can make the problem easier by doubling the integral from 0 to 2. That is, y axis symmetry means

$$\int_{-a}^{a} f(x) \, dx = 2 \int_{0}^{a} f(x) \, dx$$

Following the section on volumes of rotations, some of the students wondered if you always have a π when you do volumes by sections. These two additional examples deal with that question.

EXAMPLE 15—

Base $x^2 + y^2 = 9$. Sections perpendicular to the x axis are equilateral triangles. Find the volume.

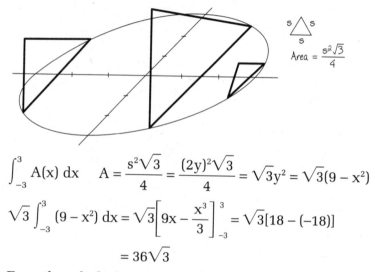

$$\int_{-3}^{3} A(x) \, dx \qquad A = \frac{s^2\sqrt{3}}{4} = \frac{(2y)^2\sqrt{3}}{4} = \sqrt{3}y^2 = \sqrt{3}(9 - x^2)$$

$$\sqrt{3}\int_{-3}^{3} (9 - x^2) \, dx = \sqrt{3}\left[9x - \frac{x^3}{3}\right]_{-3}^{3} = \sqrt{3}[18 - (-18)]$$

$$= 36\sqrt{3}$$

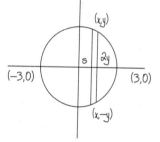

Even though the base is a circle, we are adding triangular slices. Since the area of a triangle does not involve π, neither does our volume.

EXAMPLE 16—

Base bounded by $y = x^{1/2}$, $x = 9$, x axis. Sections perpendicular to the x axis are semicircles. Find the volume.

$$\int_0^9 A(x)\,dx \qquad A = \frac{1}{2}\pi r^2 = \frac{1}{2}\pi(y/2)^2 = \frac{\pi}{8}y^2 = \frac{\pi x}{8}$$

$$\frac{\pi}{8}\int_0^9 x\,dx = \frac{\pi x^2}{16}\Big|_0^9 = \frac{81\pi}{16}$$

Since we are adding circular slices (well, to be absolutely accurate, semicircular slices), the volume does have a π in it.

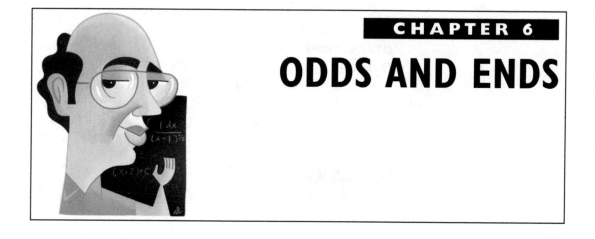

CHAPTER 6

ODDS AND ENDS

This is the chapter which covers everything we don't want to put in any of the other chapters.

DIFFERENTIALS

The first topic is differentials. One of the wonderful things calculus does is give techniques for approximations. Since we live in an imperfect world (really?), approximations are very necessary (exact answers are nicest, of course, but are not usually necessary and sometimes not attainable). Later we will get more sophisticated means of approximation. This method is usually the first.

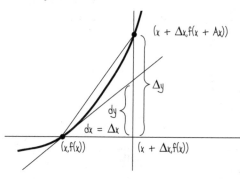

DEFINITION

Differential—dy = f'(x) dx. [y = f(x).]

Δy is the actual change; dy is the approximate change. If Δx is small when compared to x, then dy is a good approximation for Δy. In the picture Δx is large. Otherwise you couldn't see the picture.

EXAMPLE 1—

Find the differential dy if $y = f(x) = x^3 + x^2$.

$dy = f'(x) \, dx = (3x^2 + 2x) \, dx$

EXAMPLE 2—

A 9-inch steel cube is heated so that each side is increased by 1/100 of an inch. Find the actual change and the approximate change in the volume.

$$V = x^3 \qquad \Delta V = (9.01)^3 - 9^3 = 731.432701 - 729$$

$$= 2.432701$$

$$dV = V'(x) \, dx = 3x^2 \, dx = 3(9)^2(0.01) = 2.43$$

Notice the difference between ΔV and dV, 2.432701 − 2.43 = 0.002701, is very small, especially when compared to a volume of 729 cubic inches, since dx = 0.01 is small when compared to x = 9.

EXAMPLE 3—

Approximate $\sqrt[3]{63.3}$ using differentials.

We first locate the closest number that is an exact cube: x = 64. dx = 63.3 − 64 = −0.7. Computing the differential at x = 64, $y = x^{1/3}$.

$$dy = \frac{dx}{3x^{2/3}} = \frac{-0.7}{3(64)^{2/3}} = \frac{-0.7}{3(16)} = \frac{-7}{480}$$

So our approximation is

$$\sqrt[3]{63.3} = \sqrt[3]{64} + dy = 4 - 7/480 = 3\,\frac{473}{480}$$

MEAN VALUE THEOREM

The mean value theorem (MVT) is a very important
theorem in differential calculus that periodically pops
up to give us more useful information.

The mean value theorem: Let f(x) be continuous on
[(a,b)]. Let f'(x) exist on (a,b). There exists a point c
between a and b such that f'(c) = f(b) – f(a)/b – a.

Let us translate. There are no breaks between a and
b, including both ends. The derivative exists except at
possibly the left and right ends. Otherwise it is
smooth. f'(c) is the slope at c. What the heck is the
rest? If x = a, the y value is f(a). If x = b, the y value is
f(b). The two end points are (a,f(a)) and (b,f(b)). The
slope of the line joining the two end points is [f(b) –
f(a)]/(b – a).

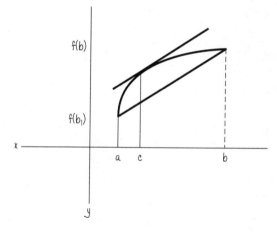

The theorem says there is at least one point between
a and b where the curve has the same slope as the line
joining the end points of the curve.

QUESTION I

Can there be more than one such point? The answer is
yes, but one and only one is guaranteed by the theorem.

QUESTION 2

Can there be such a point if the continuity or differentiability condition does not hold? The answer is yes, but there also may not be such a point. In both the cases illustrated below, there is no point on the curve where the slope is the same as the slope of the line joining the end points of the curve.

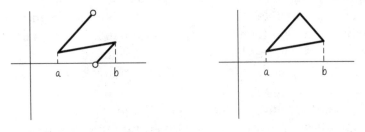

EXAMPLE 4—

Find the point c in the mean value theorem or explain why it does not exist if $f(x) = x^2 + 3x + 5$ on $[1,4]$.

Polynomials are always continuous and differentiable everywhere. The MVT holds. $f'(x) = 2x + 3$. $f'(c) = 2c + 3$. Therefore

$$2c + 3 = \frac{f(4) - f(1)}{4 - 1} = \frac{33 - 9}{3} \qquad 2c + 3 = 8 \qquad c = 2.5$$

Notice that 2.5 is between 1 and 4. Let's try others.

EXAMPLE 5—

Same as above if

$$f(x) = \frac{4}{x - 3}, \ [1,4]$$

The MVT does not hold since the function is not continuous at $x = 3$.

EXAMPLE 6—

Same as above except $f(x) = (x - 4)^{1/3}$, $[0,10]$.

f′(x) = 1/[3(x − 4)$^{2/3}$] and the derivative does not exist at x = 4. The mean value theorem again does not hold.

EXAMPLE 7—

Same as above except f(x) = x$^{1/3}$, [0,8].

   f′(x) = 1/(3x$^{2/3}$). The function is continuous. The derivative exists at every point except x = 0. The MVT holds because f′(x) does not have to exist at the end points.

$$f'(c) = \frac{1}{3c^{2/3}} = \frac{f(8) - f(0)}{8 - 0} = \frac{2}{8} = \frac{1}{4}$$

$$3c^{2/3} = 4 \qquad \text{so} \qquad c = \left(\frac{4}{3}\right)^{3/2}$$

NOTE

You alert students might have seen in Example 6 that the MVT does not hold, yet there is a point that satisfies the theorem. This is possible. If the MVT holds, such a point c is guaranteed. If the MVT doesn't hold, such a point is possible but not guaranteed.

APPROXIMATIONS, APPROXIMATIONS

In an age where computers and calculators, even those fun graphing calculators, do so many things, some things cannot be done exactly.

   We know that all quadratics can be solved using the quadratic formula. Similarly, there is a cubic formula and a quartic (fourth-degree) formula that can solve all cubics or quartics (although they are truly ugly and messy). However, in higher mathematics we can prove that most general fifth-degree equations cannot be solved. More simply, an equation like 2x = cos x cannot be solved exactly. However, we can approximate a solution very closely.

Newton's Method

Suppose we have an equation $y = f(x)$. Let's say it crosses the x axis at $x = r$. That is, the root $f(r) = 0$, but it is not exact. If we can find $f(a) < 0$ and $f(b) > 0$, then if $f(x)$ is continuous, there is a point r such that $f(r) = 0$, and r is between a and b. We can use Newton's method.

1. Let x_1 be the first approximation. Draw a tangent line at the point $(x_1, f(x_1))$ until it hits the x axis at x_2, which is usually closer to r than x_1. Continue. . . .

2. Let us give a formula using point slope.

   $$y - f(x_1) = f'(x_1)\,(x - x_1)$$

   If $f'(x_1) \neq 0$, the line is not parallel to the x axis, and the line hits the x axis at, let's say, the point $(x_2, 0)$. Substitute the point in (2).

3. $0 - f(x_1) = f'(x_1)\,(x_2 - x_1)$. We solve for x_2.

4. $x_2 = x_1 - \dfrac{f(x_1)}{f'(x_1)}$

   Repeating, we get the general formula

   $$x_{n+1} = x_n - \frac{f(x_n)}{f'(x_n)}$$

Let's do an example.

EXAMPLE 8—

Find the root of $f(x) = x^3 - x - 3$ using Newton's method.

The picture of the graphing, using the fun TI 82 calculator, looks like this:

$f(x) = x^3 - x - 3$.

Therefore $f'(x) = 3x^2 - 1$.

Newton's formula becomes

$$x_{n+1} = x_n - \frac{f(x_n)}{f'(x_n)}$$

$$= x_n - \frac{x_n^3 - x_n - 3}{3x_n^2 - 1}$$

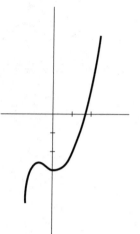

Simplifying, we get

$$x_{n+1} = \frac{2x_n^3 + 3}{3x_n^2 - 1}$$

$f(1) = -3$ and $f(2) = 3$.

Let's take $x_1 = 1.5$ for the first approximation. Probably $x_1 = 1$ or $x_1 = 2$ would work OK.

$$x_2 = \frac{2(1.5)^3 + 3}{3(1.5)^2 - 1} = 1.695652174$$

$$x_3 = \frac{2(1.695652174)^3 + 3}{3(1.695652174)^2 - 1} = 1.672080792$$

Continuuuuuing, we get $x_4 = 1.671700271$, $x_5 = 1.671699882$, $x_6 = 1.671699882$. We have reached the accuracy of our calculator, nine decimal places, probably more accurate than we would ever need.

NOTE 1
Sometimes the method doesn't work. A full study is left to other courses.

NOTE 2
When the method works, it usually works very quickly with great accuracy.

NOTE 3
This topic, as well as the other two approximations, can be done at many levels. Most appropriately, they occur in either calc I or II.

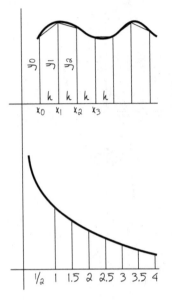

Trapezoidal Method

This method, the only one of the three that does not actually require calculus, approximates the area under the curve by trapezoids, by approximating the "top" of the region by a line.

Divide the region into n equal parts. See the trapezoids. They are standing on their heights. The area of a trapezoid is ½ h ($b_1 + b_2$). All of the h's are the same. For the first trapezoid, $b_1 = y_1$ and $b_2 = y_2$. A = ½ h ($y_1 + y_2$). For the second trapezoid, $b_1 = y_2$ and $b_2 = y_3$. A = ½ h ($y_2 + y_3$).

Notice that the lower base of the first trapezoid is the upper base of the second trapezoid. Every base is doubled except the first upper base and the last lower base. The formula is A = ½ h ($y_0 + 2y_1 + 2y_2 + \cdots + 2y_{n-1} + y_n$).

EXAMPLE 9—

Approximate $\int_1^4 \dfrac{1}{x}\, dx$ using six equal subdivisions.

The interval is of length $4 - 1 = 3$.
Six equal parts? Each h = 3/6 = ½.

$$x_0 = 1 \qquad y_0 = \frac{1}{1} = 1$$

$$x_1 = \frac{3}{2} \qquad y_1 = 1\left(\frac{3}{2}\right) = \frac{2}{3}$$

$$x_2 = 2,\ y_2 = \frac{1}{2};\ x_3 = \frac{5}{2},\ y_3 = \frac{2}{5};\ y_4 = \frac{1}{3},\ y_5 = \frac{2}{7},\ y_6 = \frac{1}{4}$$

$$A = \tfrac{1}{2}h\ (y_0 + 2y_1 + 2y_2 + 2y_3 + 2y_4 + 2y_5 + y_6)$$

$$= \frac{1}{4}\left(1 + 2\left(\frac{2}{3}\right) + 2\left(\frac{1}{2}\right) + 2\left(\frac{2}{5}\right) + 2\left(\frac{1}{3}\right) + 2\left(\frac{2}{7}\right) + \frac{1}{4}\right)$$

$$= 1.405357143$$

Parabolic Method

Another method is to approximate the region using parabolas on the top. We will isolate one of these regions. As you will see, only if n = even, an even number of intervals, will this method work.

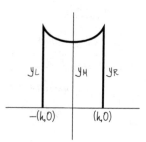

Let the parabola be given by $y = f(x) = ax^2 + bx + c$.
y at the left end is $y_L = f(-h) = ah^2 - bh + c$
y in the middle is $y_M = f(0) = c$
y at the right is $y_R = f(h) = ah^2 + bh + c$
The area of this region is

$$\int_{-h}^{h} (ax^2 + bx + c) \, dx$$

$$= ax^3/3 + bx^2/2 + cx \left[\vphantom{\int}\right]_{-h}^{h}$$

$$= ah^3/3 + bh^2/2 + ch - (-ah^3/3 + bh^2/2 - ch)$$

$$= 2ah^3/3 + 2ch$$

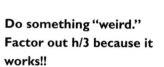

Do something "weird." Factor out h/3 because it works!!

$$= h/3(2ah^2 + 6ch)$$

$$= h/3(2ah^2 + 2ch + 4ch)$$

$$= h/3(y_L + y_R + 4y_M)$$

Just like before, the lower base of the first region is the upper base of the second region. Four times the middle never changes.

The formula issss . . . $A = h/3(y_0 + 4y_1 + 2y_2 + 4y_3 + \cdots + 4y_{n-1} + y_n)$, n even.

EXAMPLE 10—

Let's do the same example.

Approximate

$$\int_1^4 \frac{1}{x}\, dx$$

$$A = \frac{h}{3}(y_0 + 4y_1 + 2y_2 + 4y_3 + 2y_4 + 4y_5 + y_6)$$

$$A = 1/6\,(1 + 4(2/3) + 2(½) + 4(2/5) + 2(1/3) + 4(2/7) + ¼)$$

$$= 1.387698413$$

As some of you know and the rest will find out soon, the exact answer is the ln 4 = 1.386294361, approximately.

CONIC SECTIONS

Most books call the circle, parabola, ellipse, and hyperbola *conic sections* without explaining why. These curves are found by passing a plane through the truncated (cut-off) right circular cone pictured here. They are formed as follows:

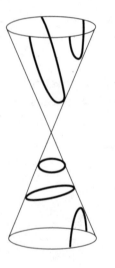

Circle—plane parallel to the top or bottom

Ellipse—plane on the top or bottom not parallel to the top or bottom but hitting all parts of the outside

Parabola—a plane parallel to an edge of the cone

Hyperbola—formed by a plane intersecting the top and bottom

DEFINITION

Circle—The set of points equidistant from a point, the center (h,k). That distance is r, the radius.

The distance formula:

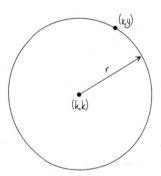

$$r = [(x - h)^2 + (y - k)^2]^{1/2}$$

Squaring, we get $(x - h)^2 + (y - k)^2 = r^2$.

EXAMPLE II—

Find the radius and center if $(x - 4)^2 + (y + \frac{1}{2})^2 = 11$. r = $11^{1/2}$, center $(4, -\frac{1}{2})$.

EXAMPLE I2—

Find the center and radius of the circle $2x^2 + 2y^2 + 8x - 16y + 6 = 0$.

   In order to do this, we have to complete the square, something we have not done since the derivation of the quadratic formula.

$2x^2 + 2y^2 + 8x - 16y + 6 = 0$

$x^2 + y^2 + 4x - 8y + 3 = 0$ **Divide by the coefficient of x^2.**

$x^2 + 4x + y^2 - 8y = -3$ **Group the x terms and y terms; get the constant to the other side.**

$x^2 + 4x + 4 + y^2 - 8y + 16 = -3 + 4 + 16$ **Take half of 4, square it, add it to both sides, and take half of −8, square it, add it to both sides.**

$(x + 2)^2 + (y - 4)^2 = 17$ **Factor into perfect squares (that was the idea) and add the terms on the right.**

The center is $(-2,4)$, r = $17^{1/2}$.

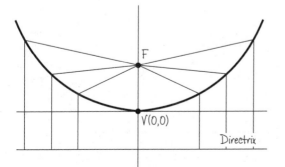

For the parabola, ellipse, and hyperbola, it is essential to relate the equation to the picture. If you do, these curves are very simple.

DEFINITION

Parabola—The set of all points equidistant from a point, called a *focus,* and a line, called a *directrix*. Point V is the vertex, equidistant from the focus and directrix and closest to the directrix and to the focus.

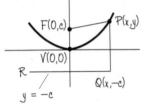

Let's do this development algebraically. Let the vertex be at (0,0). The focus is (0,c). The directrix is y = −c. Let (x,y) be any point on the parabola. The definition of a parabola says FP = PQ. Just like before, everything on PQ has the same x value and everything on RQ has the same y value. The coordinates of Q are (x,−c). Since the x values are the same, the length of PQ = y − (−c). Using the distance formula to get FP and setting it equal to FP, we get $((x − 0)^2 + (y − c)^2)^{1/2} = y + c$. Squaring, we get $x^2 + y^2 − 2cy + c^2 = y^2 + 2cy + c^2$. Simplifying, we get $x^2 = 4cy$.

We will make a small chart relating the vertex, focus, directrix, equation, and picture.

	Vertex	Focus	Directrix	Equation	Picture	Comment
1.	(0,0)	(0,c)	y = −c	$x^2 = 4cy$		The original derivation
2.	(0,0)	(0,−c)	y = c	$x^2 = −4cy$		y replaced by −y
3.	(0,0)	(c,0)	x = −c	$y^2 = 4cx$		x,y interchange in 1
4.	(0,0)	(−c,0)	x = c	$y^2 = −4cx$		x replaced by −x in 3

If you relate the picture to the original equation, the
sketching will be easy.

EXAMPLE 13—

Given $y^2 = -7x$. Sketch. Label vertex, focus, directrix.
  From the chart, we know the sketch is picture 4.
Now let $4c = 7$ (ignore the minus sign). $c = 7/4$. The
vertex is (0,0). The focus is (−7/4,0), because it is on
the x axis to the left of the origin. The directrix is y =
7/4; y, a vertical line, = +7/4 because it is to the right of
the origin.

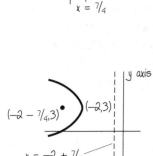

EXAMPLE 14—

Sketch $(y - 3)^2 = -7(x + 2)$.
  To understand the following, we need only note the
difference between $x^2 + y^2 = 25$ and $(x - 3)^2 + (y + 6)^2 =$
25. Has the shape changed? No. Has the radius
changed? No. What has changed? The center. Instead
of being at the point (0,0), the center is at the point
(3,−6).
  In the case of the parabola, what has changed is the
vertex. Instead of being at the point (0,0), the vertex
is at the point (−2,3). The shape is the same. 4c still
equals 7. So c = 7/4. The focus now becomes (−2 −
7/4,3), 7/4 to the left of the vertex (−7/4 from the x
coordinate). The directrix is x = −2 + 7/4.

EXAMPLE 15—

Sketch the parabola $2x^2 + 8x + 6y + 10 = 0$.

$2x^2 + 8x + 6y + 10 = 0$ **Original.**

$x^2 + 4x + 3y + 5 = 0$ **Divide through by the
coefficient of the squared
variable.**

On one side, get all the terms that have the squared letter; everything else goes to the other side.	$x^2 + 4x = -3y - 5$
Complete the square; add to both sides.	$x^2 + 4x + 4 = -3y - 5 + 4$
Factor and simplify.	$(x + 2)^2 = -3y - 1$
This is weird. No matter what the coefficient on the right side, factor the whole coefficient out, even if there is a fraction in the parentheses.	$(x + 2)^2 = -3(y + 1/3)$

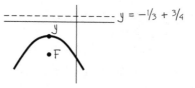

Sketch v(−2,−1/3), shape 2, ∩. 4c = 3, c = 3/4. F(−2,−1/3,−3/4). Directrix y = −1/3 + 3/4.

We will now look at the ellipse. Algebraically, the *ellipse* is defined as $PF_1 + PF_2 = 2a$, where $2a > 2c$, the distance between F_1 and F_2. In words, given two points, F_1 and F_2, two foci. If we find all points P, such that if we go from F_1 to P and then from P to F_2, add those two distances together, and we will always get the same number, 2a, where a will be determined later; we will get an ellipse.

I know you would desperately like to know how to draw an ellipse. This is how. Take a nonelastic string. Attach both ends with thumbtacks to the table. Take the point of a pencil and stretch the string as far as it will go. Go 360 degrees. You will trace out an ellipse.

Some of you have seen the equation for an ellipse, but few of you have seen the derivation. It is an excellent algebraic exercise for you to try. You will see there is a lot that goes into a rather simple equation.

$PF_1 + PF_2 = 2a$

$\sqrt{[x - (-c)]^2 + (y - 0)^2} + \sqrt{(x - c)^2 + (y - 0)^2} = 2a$

$[\sqrt{(x + c)^2 + y^2}]^2 = [2a - \sqrt{(x - c)^2 + y^2}]^2$

$x^2 + 2cx + c^2 + y^2$

$\qquad = 4a^2 + x^2 - 2cx + c^2 + y^2 - 4a\sqrt{(x - c)^2 + y^2}$

$4cx - 4a^2 = -4a\sqrt{(x - c)^2 + y^2}$

$P(x,y)$

$F_1(-c,0) \quad F_2(c,0)$

Combine like terms; isolate the radical.

$(cx - a^2)^2 = [-a\sqrt{(x - c)^2 + y^2}]^2$

$a^4 - 2a^2cx + c^2x^2 = a^2(x^2 - 2cx + c^2 + y^2)$

$a^4 - a^2c^2 = a^2x^2 - c^2x^2 + a^2y^2$

$\dfrac{(a^2 - c^2)x^2 + a^2y^2}{(a^2 - c^2)a^2 \quad a^2(a^2 - c^2)} = \dfrac{a^2(a^2 - c^2)}{a^2(a^2 - c^2)}$

Reverse sides; take out common factors.

Divide on both sides by $a^2(a^2 - c^2)$.

Let $a^2 - c^2 = b^2$.

$\dfrac{x^2}{a^2} + \dfrac{y^2}{a^2 - c^2} = 1$

Finally we get

$\dfrac{x^2}{a^2} + \dfrac{y^2}{b^2} = 1$

Whew!

We are still not finished. What is a and what is b? Let's investigate.

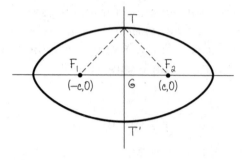

Since T is any point on the ellipse, $F_1T + TF_2 = 2a$. By symmetry, $F_1T = TF_2$. So $F_1T = a$. Since $a^2 - c^2 = b^2$, $GT = b$. The coordinates of T are $(0,b)$, and the coordinates of T' are $(0,-b)$.

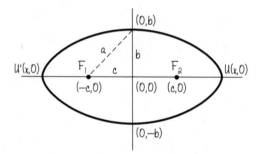

We would like to find the coordinates of U and we have used up the letters a, b, and c. Oh well, let's see what happens. $F_2U + UF_1 = 2a$. $F_2U = x - c$. $UF_1 = x + c$. $x + c + x - c = 2a$. So $x = a$. The coordinates of U are $(a,0)$. The coordinates of U' are $(-a,0)$.

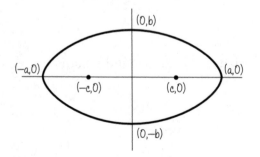

c = half the distance between the foci. b = length of the *semiminor axis* ("semi" means half, "minor" means smaller, "axis" means line). a = length of the *semimajor axis* = distance from a focus to a minor vertex. $(\pm a,0)$ are the *major vertices*. $(0,\pm b)$ are the *minor vertices* or *co-vertices*. $(\pm c,0)$ are the foci.

Although the derivation is very long, sketching should be short.

EXAMPLE 16—

Sketch $x^2/11 + y^2/8 = 1$.

In the case of an ellipse, the longer axis is indicated by which number is larger under x^2 or y^2. That term is a^2. (Try not to remember a or b—remember the picture.) This ellipse is longer in the x direction.

Letting $y = 0$, we get the major vertices $(\pm\sqrt{11},0)$. Letting $x = 0$, we get the minor vertices $(0,\pm\sqrt{8})$. c = $\sqrt{11 - 8}$. The foci are $(\pm\sqrt{3},0)$. The sketch is below.

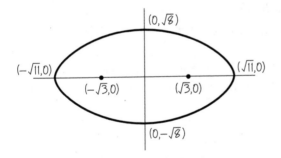

EXAMPLE 17—

Sketch $x^2/5 + y^2/26 = 1$.

Major vertices: $(0,\pm\sqrt{26})$. Minor vertices: $(\pm\sqrt{5},0)$. Foci: $(0,\pm\sqrt{21})$—foci are always on the longer axis.

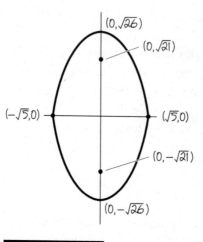

EXAMPLE 18—

$$\frac{(x-6)^2}{11} + \frac{(y+4)^2}{8} = 1$$

This is the same as Example 12 except that the center is no longer at the point (0,0). It is moved to the point (6,–4). Now the major vertices are $(6 \pm \sqrt{11}, -4)$. The minor vertices are $(6, -4 \pm \sqrt{8})$. The foci are $(6 \pm \sqrt{3}, -4)$.

Weird numbers are intentionally chosen so that you know exactly where the numbers come from.

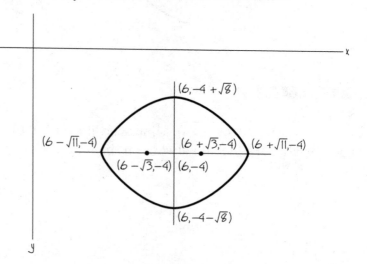

EXAMPLE 18 REVISITED—

$(x - 6)^2/11 + (y + 4)^2/8 = 1$

Just a few months after the book went to press, I realized a second way to find the vertices that made it much clearer, but my editor wouldn't change the book. So now I finally have a chance to show you.

The center is again at the point (6,–4). The vertices are directly east-west and north-south of the point (6,–4). All points east-west of (6,–4) must have the same y value, y = –4. Sooo, letting y = –4, we get

$$\frac{(x - 6)^2}{11} + \frac{(-4 + 4)^2}{8} = 1$$

$$\frac{(x - 6)^2}{11} + 0 = 1$$

$$(x - 6)^2 = 11$$

$$x - 6 = \pm 11^{1/2}$$

Therefore $x = 6 \pm 11^{1/2}$ and the major vertices are $(6 \pm 11^{1/2}, -4)$.

Points north-south of (6,–4) have the same x value, x = 6.

$$\frac{(6 - 6)^2}{11} + \frac{(y + 4)^2}{8} = 1$$

$$0 + \frac{(y + 4)^2}{8} = 1$$

$$(y + 4)^2 = 8$$

$$y + 4 = \pm 8^{1/2}$$

Therefore $y = -4 \pm 8^{1/2}$ and the minor vertices are $(6, -4 \pm 8^{1/2})$. For the same reasons the foci, always on the major axis, are $(6 \pm 3^{1/2}, -4)$. The sketch is, of course, the same.

EXAMPLE 19—

Sketch and discuss $4x^2 + 5y^2 + 30y - 40x + 45 = 0$.

Here we must complete the square in a slightly different manner.

$$4x^2 + 5y^2 + 30y - 40x + 45 = 0$$

$$4x^2 - 40x + 5y^2 + 30y = -45$$

$$4\left(x^2 - 10x + \left(\frac{-10}{2}\right)^2\right) + 5\left[y^2 + 6y + \left(\frac{6}{2}\right)^2\right]$$

$$= -45 + 4\left(\frac{-10}{2}\right)^2 + 5\left(\frac{6}{2}\right)^2$$

$$\frac{4(x-5)^2}{100} + \frac{5(y+3)^2}{100} = \frac{100}{100}$$

$$\frac{(x-5)^2}{25} + \frac{(y+3)^2}{20} = 1$$

Center: $(5,-3)$. Vertices: $(5 \pm \sqrt{25}, -3)$, $(5, -3 \pm \sqrt{20})$.
$c = \sqrt{25 - 20} = \sqrt{5}$. Foci: $(5 \pm \sqrt{5}, -3)$.

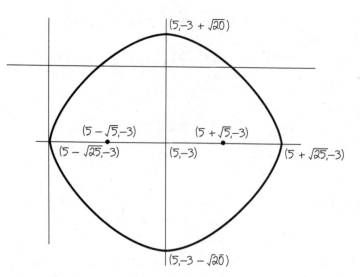

Of course you should use 5 instead of $\sqrt{25}$. I leave the $\sqrt{25}$ so you know where it came from.

The definition of the *hyperbola* is $F_1P - PF_2 = 2a$, where the foci are $(\pm c,0)$. The derivation is exactly the same as for an ellipse. Once is enough!! The equation we get is $x^2/a^2 - y^2/b^2 = 1$, where $a^2 + b^2 = c^2$. $(\pm a,0)$ are called the *transverse vertices.* The hyperbola has asymptotes $y = \pm(b/a)x$.

NOTE I

The shape of a hyperbola is determined by the location of the minus sign, not which number is larger under the x^2 or y^2.

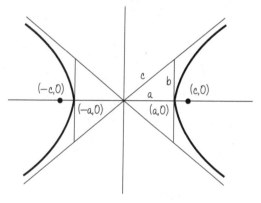

NOTE 2

In the case of the asymptote, the slope of the line b/a is the square root of the number under the y^2 divided by the square root of the term under the x^2 term.

EXAMPLE 20—

Sketch and label $x^2/7 - y^2/11 = 1$.

Transverse vertices: $y = 0$, $x = \pm\sqrt{7}$ $(\pm\sqrt{7},0)$. Note that if $x = 0$, $y = \pm\sqrt{-11}$, which are imaginary. The curve does not hit the y axis. $c = \sqrt{7 + 11}$. Foci: $(\pm\sqrt{18},0)$. $y = \pm(\sqrt{11}/\sqrt{7})x$.

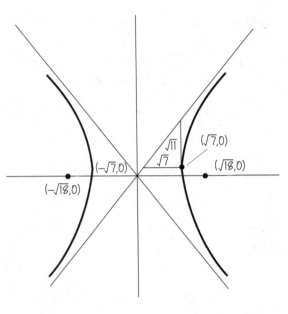

EXAMPLE 21—

Sketch and discuss $y^2/5 - x^2/17 = 1$.

Transverse vertices: $(0, \pm\sqrt{5})$. Foci: $(0, \pm\sqrt{22})$.
Asymptotes: $y = \pm(\sqrt{5}/\sqrt{17})x$.

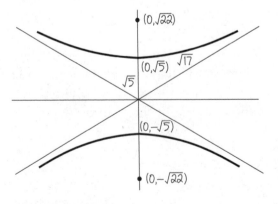

EXAMPLE 22—

The same form

$$\frac{(y - 6)^2}{5} - \frac{(x + 7)^2}{17} = 1$$

This is the same as the above sketch except that the "center of the hyperbola," the place where the asymptotes cross, is no longer (0,0). It is now (−7,6). Transverse vertices: (−7,6 ± $\sqrt{5}$). Foci: (−7,6 ± $\sqrt{22}$). Asymptotes: y − 6 = ±($\sqrt{5}/\sqrt{17}$)(x + 7).

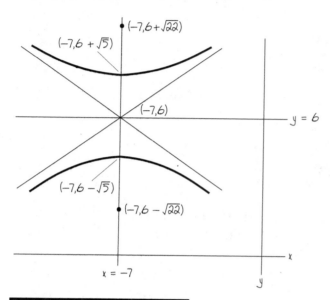

EXAMPLE 22 REVISITED—

$(y − 6)^2/5 − (x + 7)^2/17 = 1$

Just like with the ellipse, let's clarify the location of the vertices. In this case, the vertices occur north-south of the center (−7,6). So the x value of the vertices is the same as the center, x = −7. Substituting, we get

$$\frac{(y − 6)^2}{5} − \frac{(−7 + 7)^2}{17} = 1$$

$(y − 6)^2 = 5$

$y = 6 ± 5^{1/2}$

The vertices are (−7,6 ± $5^{1/2}$).

For the same reason the foci, always on the axis through the vertices, are (−7,6 ± $22^{1/2}$). The asymptotes and the sketch are the same!!

EXAMPLE 23—

Sketch and discuss $25x^2 - 4y^2 + 50x - 12y + 116 = 0$.

For the last time we will complete the square, again a little bit differently.

$$25x^2 - 4y^2 + 50x - 12y + 116 = 0$$

$$25x^2 + 50x - 4y^2 - 12y = -116$$

$$25[x^2 + 2x + (2/2)^2] - 4[y^2 + 3y + (3/2)^2]$$

$$= -116 + 25(2/2)^2 - 4(3/2)^2$$

$$\frac{25(x+1)^2}{-100} - \frac{4(y+3/2)^2}{-100} = \frac{-100}{-100}$$

$$\frac{(y+3/2)^2}{25} - \frac{(x+1)^2}{4} = 1$$

Center: $(-1,-3/2)$. Vertices: $(-1,-3/2 \pm \sqrt{25})$. Foci: $(-1,-3/2 \pm \sqrt{29})$. Asymptotes: $y + 3/2 = \pm(\sqrt{25}/\sqrt{4})(x+1)$.

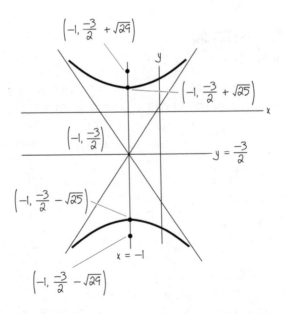

Sometimes we have a puzzle. Given some information, can we find the equation? You must *always draw
the picture* and relate the picture to its equation.

EXAMPLE 24—

Find the equation of the parabola with focus (1,3),
directrix $x = 11$.

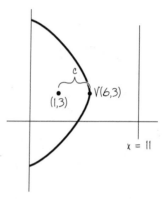

Drawing F and the directrix, the picture must be the
following picture. The vertex is halfway between the x
numbers. So $x = (11 + 1)/2 = 6$.

V(6,3). c = the distance between V and F = 5. The
equation is $(y - 3)^2 = -4c(x - 6) = -20(x - 6)$. Remember, the minus sign is from the shape and c is always
positive for these problems.

EXAMPLE 25—

Vertices (2,3) and (12,3) and one focus (11,3). Find the
equation of the ellipse.

Two vertices give the center, $((12 + 2)/2,3) = (7,3)$.
F(11,3). $(x - 7)^2/a^2 + (y - 3)^2/b^2 = 1$. $a = 12 - 7 = 5$. $c =$
$11 - 7 = 4$. $a^2 - b^2 = c^2$. $5^2 - b^2 = 4^2$. $b^2 = 9$ (no need for
b). $(x - 7)^2/25 + (y - 3)^2/9 = 1$.

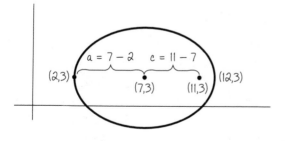

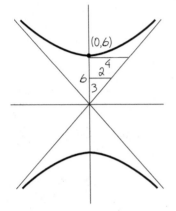

EXAMPLE 26—

Find the equation of the hyperbola with vertices $(0, \pm 6)$ and asymptotes $y = \pm(3/2)x$.

$V(0, \pm 6)$ says the center is $(0,0)$ and its shape is $y^2/36 - x^2/a^2 = 1$. The slope of the asymptotes is $3/2 =$ square root of the number under y^2 over the square root of the number under the x^2 term. So $3/2 = 6/?$. So $? = 4$. So $a^2 = 4^2 = 16$. The equation is $y^2/36 - x^2/16 = 1$.

This kind of question is shorter in length, but it does take practice. So practice!!!

L'HOPITAL'S RULE

Unless I can think of something else before this book gets printed, this is the last topic and one of the nicest.

L'Hopital's rule states that if $\displaystyle\lim_{x \to a} \frac{f(x)}{g(x)} = \frac{0}{0}$ or $\frac{\infty}{\infty}$

where a can be any finite number, plus infinity, or minus infinity, and if

$$\lim_{x \to a} \frac{f'(x)}{g'(x)} = L$$

then

$$\lim_{x \to a} \frac{f(x)}{g(x)} = L$$

Simply put, if the original limit is indeterminate, simply take the derivative of the top and divide it by the derivative of the bottom, *not* the quotient rule. If this gives a limit L (or does not exist), then the original limit is the same.

EXAMPLE 27—

$$\lim_{x \to 0} \frac{2x}{x^2 + 1} = \frac{0}{1} = 0$$

Using L'Hopital's rule is unnecessary and wrong!

EXAMPLE 28—

$$\lim_{x \to \infty} \frac{x^2 + 1}{1/x} = \lim_{x \to \infty} (x^3 + x)$$

which does not exist—same comment as in the example.

EXAMPLE 29—

$$\lim_{x \to 0} \frac{\sin x}{x} = \frac{0}{0}$$

Hooray! L'Hopital's rule applies. Taking the derivative of the top over the derivative of the bottom, we get

$$\lim_{x \to 0} \frac{\cos x}{1} = \frac{1}{1} = 1$$

So

$$\lim_{x \to 0} \frac{\sin x}{x} \text{ also} = 1$$

EXAMPLE 30—

$$\lim_{x \to \infty} \frac{x^2 - 4x}{3 - 5x^2} = \frac{\infty}{\infty}$$

L'Hopital's rule applies. Taking derivatives of the top and bottom, we get $(2x - 4)/(-10x)$, again going to ∞/∞. Use our rule twice. The second time we get $2/-10 = -1/5$. So our original limit was also $-1/5$.

EXAMPLE 31—

$$\lim_{x \to 0} (1/\sin x - 1/x) = \infty - \infty$$

We must do something algebraic so our wonderful rule can be used. In this case, we simply add the fractions together.

$$\lim_{x \to 0} (1/\sin x - 1/x) = \lim_{x \to 0} \frac{x - \sin x}{x \sin x} = \frac{0}{0}$$

L'Hopital's rule once gives us

$$\lim_{x \to 0} \frac{1 - \cos x}{x \cos x + \sin x} = \frac{0}{0}$$

Again, we get $\lim_{x \to 0} \dfrac{\sin x}{2 \cos x - x \sin x} = \dfrac{0}{2} = 0$

So the original $\lim_{x \to 0} (1/\sin x - 1/x) = 0$.

Sometimes when you take the limit, you get $\infty \times 0$. In that case you must do something algebraic to get $0/0$ or ∞/∞.

EXAMPLE 32—

$$\lim_{x \to \infty} [x \sin (1/x)]$$

Limit is $\infty \times 0$.

We rewrite $x \sin (1/x)$ as

$$\frac{\sin (1/x)}{1/x}$$

Now the limit is $0/0$, and we can apply L'Hopital's rule.

$$\frac{[\sin(1/x)]'}{(1/x)'} = \frac{\cos(1/x)(-1/x^2)}{(-1/x^2)} = \cos (1/x)$$

Since

$$\lim_{x \to \infty} \cos (1/x) = 1$$

$$\lim_{x \to \infty} [x \sin (1/x)] = 1$$

Well, that's all for now. The *Calc II* and *Calc III* will help you next. If you need algebra or trig review, get the *Precalc*.

ABOUT BOB MILLER...
IN HIS OWN WORDS

I received my B.S. and M.S. in math from Brooklyn Poly, now Polytechnic University. After my first class, which I taught as a substitute for a full professor, one student told another upon leaving the room that "at least now we have someone who can teach the stuff." I was forever hooked on teaching. Since then I have taught at Westfield State College, Westfield, Massachusetts; Rutgers; and the City College of New York, where I've been for the last 28½ years. No matter how bad I feel before class, I always feel great after I start teaching. I especially like to teach precalc and calc, and I am always delighted when a student tells me that he or she has always hated math before and could never learn it, but that taking a class with me has made math understandable and even enjoyable. I have a fantastic wife, Marlene; a wonderful daughter, Sheryl; a terrific son, Eric; and a great son-in-law, Glenn. The newest member of our family is my adorable, brilliant granddaughter Kira Lynn, eight days old as of this writing. My hobbies are golf, bowling, bridge, and crossword puzzles. Someday I hope a publisher will allow me to publish the ultimate high school text and the ultimate calculus text so our country can remain number one forever.

To me, teaching math always is a great joy. I hope I can give some of this joy to you.

INDEX